AF719207

Understanding Sea Anchors and Drogues

BY EARL R. HINZ

The Offshore Log, 1968

Sail Before Sunset, 1979

Landfalls of Paradise: The Guide to Pacific Islands, 1980

The Complete Book of Anchoring and Mooring, 1986

Understanding Sea Anchors and Drogues, 1987

Understanding Sea Anchors and Drogues

BY EARL R. HINZ

With drawings by RICHARD R. RHODES

Cornell Maritime Press

CENTREVILLE, MARYLAND

Copyright © 1987 by Cornell Maritime Press, Inc.

All rights reserved. No part of this book may be used or reproduced in any manner whatsoever without written permission except in the case of brief quotations embodied in critical articles and reviews. For information, address Cornell Maritime Press, Inc., Centreville, Maryland 21617.

Library of Congress Cataloging in Publication Data
Hinz, Earl R.
Understanding sea anchors and drogues.
Bibliography: p.
Includes index.
1. Anchors, Sea. I. Title. II. Title: Drogues.
VM791.H56 1987 623.8′62 87-47738
ISBN 0-87033-375-5

Manufactured in the United States of America

First edition

Contents

Acknowledgments

A book of this sort is certainly not the work of one person. It draws on the observations of many people over many years. Among those who have made their contributions to this assessment of sea anchors and drogues are Captain Voss, Fritz Fenger, Robert Manry, the Casanovas, Eric Hiscock, Frank Robb, Harry Pidgeon, Adlard Coles, Ross Norgrove, Robin Knox-Johnston, William Van Dorn, the Pardeys, Bernard Moitessier, and Bob Griffith. Through their experiences, and writings by and about them, the boating world has learned much about seamanship under adverse conditions. And we should not forget the many crewmen who participated in the disastrous Fastnet Race of 1979 and proved the need for improved survival techniques.

The author would also like to recognize contributions made by Dan Shewmon and Donald Jordan in their articles for *Cruising World* and *Sail* magazines, respectively. These lively forums have brought forth most of the problems of, and many solutions to, storm survival.

I would also like to thank two manufacturers for unstinting cooperation in sharing their development experiences on drag devices—Dan Shewmon of Shewmon, Inc., and Skip Hathaway of Hathaway, Reiser & Raymond sailmakers. They have been on the front line of evolution of a new generation of survival gear.

Finally, when it got down to penning an assessment of these controversial devices, I had the help of well-known boating people who also are authors in their own right. John Rousmaniere took the first cut at the manuscript and added considerable emphasis to the overall problem based on his personal experiences in the 1979 Fastnet Race as well as his extensive analysis of the race documented so well in his book *Fastnet, Force 10*.

Longtime working associates Freeman Pittman, technical editor of *Sail*, and Bob Vollmer, formerly technical editor of *Sea*, helped me over the editorial hurdles with early blue-penciling of the manuscript. If the writing makes sense, credit them; if it doesn't, then complain to me. It was not an easy subject to expound on because of the bewildering variety of opinions held by experts in blue water sailing.

Finally, I again had the talents of Dick Rhodes for the illustrations. Dick has been close to the sea as well as being a marine illustrator of note. He has drawn the lines for the square-rigged *Falls of Clyde* and the *Hokule'a* (the Polynesian voyaging canoe) and has illustrated many Polynesian canoe books and stories. Recently he has devoted considerable effort to a new line of catamarans for racing and charter work. But one of his most exciting ventures has been as a crewman on *Hokule'a,* which over the past three years has sailed from Hawaii to New Zealand and return—the old Polynesian way.

Let me conclude by quoting the old sailing cliche: "May your winds be fair and your seas smooth." And if they aren't, read on; you may just find a way to make it through the storm.

Understanding Sea Anchors and Drogues

ONE

Sea Anchor or Drogue?

The use of drag devices at sea is fraught with confusion, contradiction, and controversy. Confusion over what the difference is between a sea anchor and a drogue; contradiction on what constitutes a proper design; and controversy on how and when to use sea anchors and drogues or whether to use them at all. They have a place in the equipment inventory of recreational and working blue water boats but first we have to take a hard look at what has gone before that has caused this indeterminate state of affairs.

Sea anchors have been around a long time, indicating that early seafarers also felt the need to "stop" the boat for crew comfort and to keep the bow into the wind for survival. There is some evidence that early Polynesian seafarers (circa 1200 AD) used a form of sea anchor on their double-hulled voyaging canoes. This was done by attaching stone ground anchors to sennit ropes and lowering them from the bows of the canoe hulls. Supposedly this brought the bows into the wind so that the shallow-draft canoes could ride out the storm with more ease. One can safely assume there was a considerable amount of hull draft but, since the canoes were steered with long demountable oars, there was no tendency to break the rudder in a rearward movement of the canoe.

Webster's dictionary notes that the first use of the term "sea anchor" in recorded English writing occurred in 1769. The word "drogue," however, did not appear in English writings until 1875 and it is believed to have started as a variant of the word "drag."

The first well-documented account of using sea anchors and drogues came from Captain John Voss's book on his voyages in the Pacific in a converted British Columbia Indian canoe (Voss 1949). On several occasions Voss found it advantageous to rig a conical sea anchor from the bow of *Tilikum* to ride out a storm. There are several features of *Tilikum* which made it a suitable vessel for sea anchoring. First, it was a narrow-beamed boat of shallow draft. Second, it did not have great windage forward or a large keel aft that would have tended to turn it broadside to the wind and waves. Lastly, it had a removeable rudder, at least it was removeable after Voss found out on two occasions that the backward motion of the boat could break the rudder. On the third sea anchor deployment, Voss unshipped the rudder and there was no further trouble.

Even then *Tilikum* did not lie perfectly head to sea but held a position about 5 points (approximately 55°) off the wind. Voss then added a leg-o'-mutton spanker sheeted in flat and cut the riding angle down one-half. Voss claims to have "weathered sixteen gales without shipping so much as a bucket of water at one time."

That same drag device that performed so successfully as a sea anchor was also used as a drogue in crossing the dangerous bar at Melbourne,

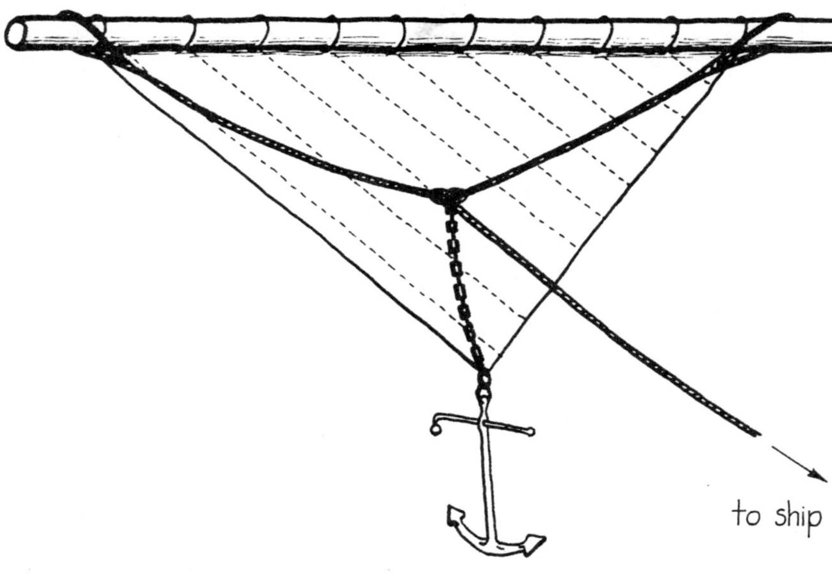

The drift sail. The earliest attempts at sea anchoring in the western world took place in the 17th century with what was then called a drift sail. It was made from a spar with sail attached. An anchor was hung from the clew of the sail with a length of chain connecting the clew to a rope bridle coming from the other corners. The hawser was knotted to the three-part bridle and the bitter end attached to the ship. Adapted from *The Mariner's Dictionary*.

Australia, and later in a demonstration of towing a boat across a bar at Whanganui, New Zealand.

Specifications for building the Voss sea anchors are given in his book and two examples are illustrated in Figure 1.

Anyone who was interested in sea anchors (or drogues) continued to use the Voss conical device until 1932 when Fritz Fenger put his analytical and mechanical mind to work in developing a new concept. Out of his experiments came the "double plank drogue." Fenger's design (*Yachting*, January 1968) was sturdy, heavy, and awkward to stow, but it worked well as either a sea anchor or a drogue—so it is said. Figure 2 is a design worked up for a boat with a 35-foot waterline. The weight of this device is approximately 55 pounds in air.

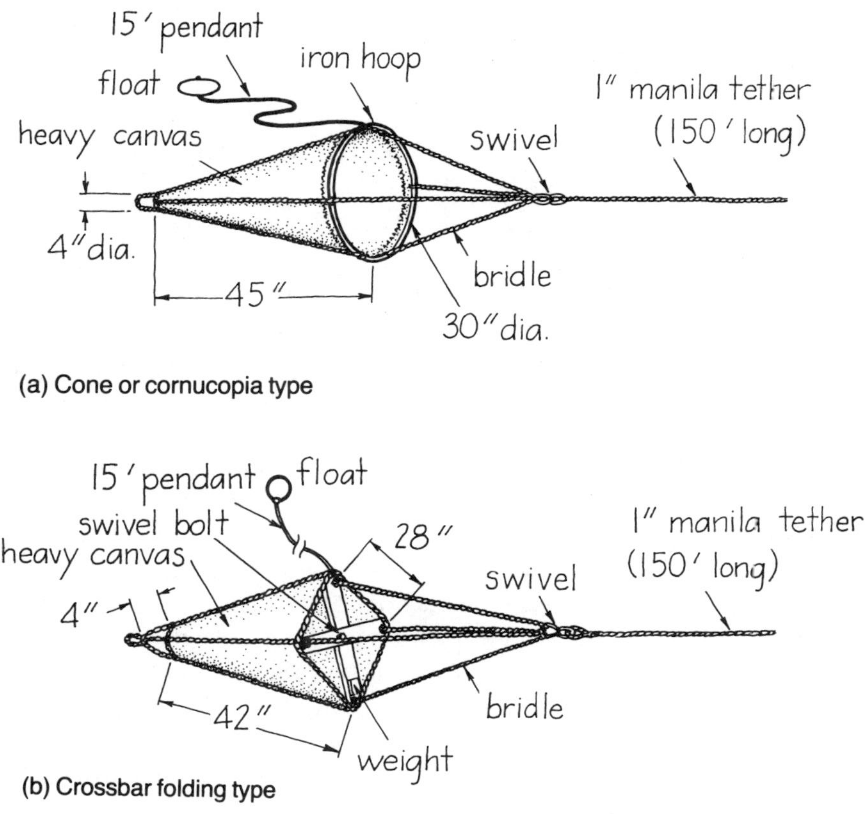

(a) Cone or cornucopia type

(b) Crossbar folding type

Figure 1. Sea anchors according to Voss (dimensions given for a deep keel boat with 30-foot waterline).

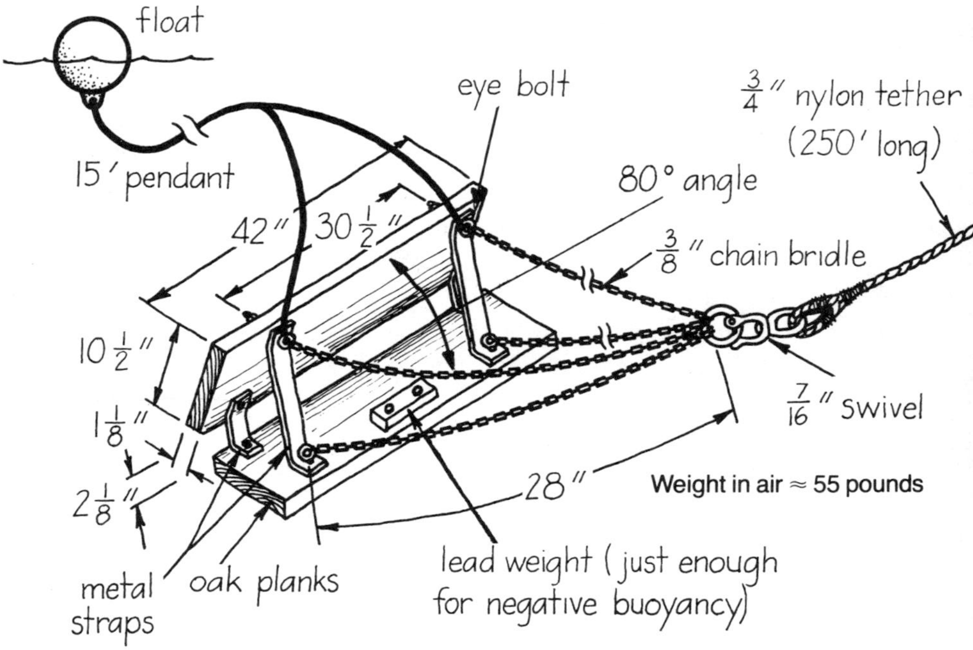

Figure 2. Fenger-type drogue for 35-foot-waterline boat.

Apparently sea anchors continued to be carried by some blue water boats but little was written about them. Most cruisers still did not have faith in them although many kept them on board just in case everything else failed. Next to use one was Robert Manry in 1965 during a gale while crossing the Atlantic in the 13½-foot-long *Tinkerbelle.* His only problem was that the boat would not lie bow to the wind until he had unshipped the transom-mounted rudder. *Tinkerbelle* then rode comfortably head to the wind (Manry 1966).

Recent proponents of a sea anchor are John and Joan Casanova who are devotees of the parachute sea anchor. They have successfully used what they term the "parachute anchoring system" on their trimaran *Tortuga Too. Tortuga Too* rode to a parachute sea anchor many times when gales set in during her 33,000-mile circumnavigation. John says, "Never wait until the last minute to put out the chute. Watch your bad weather signs. When you become nervous and a bit concerned, it is time to use the parachute and go below; relax if you can." (Casanova et al. 1982)

But not everybody has the same feelings about sea anchors and drogues. Reading through the literature of cruising one finds that Eric

Hiscock has used one (a drogue) and believes it is an "essential piece of gear . . . enabling [a boat] to ride out the worst kind of weather in safety and by reducing her drift to a minimum, keeping her off a lee shore." He also found out that a good drogue has to be stronger than the traditional canvas sea anchor since his was essentially shredded by the time it was hauled aboard (Hiscock 1956).

Frank Robb has also had drogue experience and says that they "have many vices and only one virtue—they check drift." (*Sea,* January 1979)

And Hal Roth, a blue water sailor of much experience, says, "The theory of a yachtsman's sea anchor is . . . to hold the ship's head to the wind and expose the strongest and most streamlined part of the vessel to the storm. . . . This is a nice fairy tale, which, unfortunately, is nonsense."

Harry Pidgeon carried a sea anchor made out of some "boards" on his shakedown cruise between Los Angeles and Hawaii in 1921 but there is no evidence that he ever used his contraption. After that cruise he made a new sea anchor out of canvas which he took along on his world cruise.

Pidgeon had no occasion to use his new sea anchor until early in 1924, when he departed Durban, South Africa. The Agulhas current carried him steadily south but soon headwinds turned into a westerly gale on the Agulhas bank and he hove-to under reefed mizzen and storm jib. The combers of the wild sea would drive the bow off and slew *Islander* almost beam-on. After the storm jib split, he put out the sea anchor, but alas, it apparently had been designed after Voss's recommendations and *Islander* was in no way a *Tilikum.* Pidgeon says, "It [the sea anchor] appeared to be too light to sink to the proper depth and was not large enough." So Pidgeon set about repairing the split storm jib and later rode out the rest of the storm hove-to (Pidgeon 1954).

Ross Norgrove tried a Fenger drogue once and proved the difference between a sea anchor and a drogue. In his book *The Cruising Life* he describes using a 5-foot-span Fenger-type drogue streamed on 100 fathoms of ⅝-inch-diameter rope from the bow of *White Squall* in a Roaring Forties gale. The response of the boat was to lie beam-on in the trough. As she would rise up on the face of one wave, the bow would start to swing downwind and the shoulder of the boat would take the punch of the next wave. He felt a lesser-built boat would not have survived the pounding.

Nevertheless, enough was enough and even if the boat was taking these shoulder punches, the crew could not. So Norgrove brought 50 feet of the sea anchor warp outside of everything and made it fast to the aft Samson post. When the warp was freed from the forward Samson post,

White Squall quickly turned end for end and took a more comfortable position with the sea. His final observation was, "Upon reflection, I would say that without that wooden contraption over the stern, or something equally as good, we would not have been anywhere near as safe."

Norgrove later encountered a true sea anchor in use on a 50-foot ex-Bluenose schooner. It was a 25-foot-diameter parachute that was capable of holding the schooner's nose into a 25-knot wind. While he thought this was the answer to a seaman's prayer, he still had reservations about holding the boat too securely when it may have been better turning and running with a drogue (Norgrove 1980).

Much of the confusion and contradiction on sea anchors and drogues can be traced back to John Voss's book *The Venturesome Voyages of Captain Voss,* in which he relates his experiences with sea anchors on *Tilikum.* He used the same drag device over the bow for a sea anchor as he used over the stern to run a bar and he always called it a sea anchor (Voss 1949).

Henderson, writing in *Sea Sense* (1979), makes a nominal differentiation between the drag devices:

> Sometimes a distinction is made between a drogue and a sea anchor but essentially each is a floating anchor designed to slow a vessel's drift and to hold her in a desired attitude. We could differentiate between the two to some extent by saying that the sea anchor is intended to reduce drift to a bare minimum, while the drogue is merely a drag to lessen drift.

Another piece of contemporary literature defines a sea anchor as follows:

> SEA ANCHOR: Any object or collection of objects streamed to windward to slow down the leeward rush of a vessel in heavy weather. A true sea anchor is a canvas open-ended cone which is streamed on the end of a line from a bridle. Under practical conditions the sea anchor tends to be insufficient or it readily carries away. One of the best alternatives is a collection of floorboards, fenders and tires lashed together and streamed. Long bights of rope and chain with nothing attached can also serve to slow the vessel down.

Even Webster's dictionary does not offer a clear, concise differentiation between the two drag devices. The *Ninth New Collegiate Dictionary* published by Merriam-Webster in 1983 says:

> SEA ANCHOR: a drag typically of canvas thrown overboard to retard the drifting of a ship or seaplane and to keep its head to the wind.
> DROGUE: 1: SEA ANCHOR 2a: a cylindrical or funnel-shaped device

towed as a target by an airplane b: a small parachute for stabilizing or decelerating something (as an astronaut's capsule) or for pulling a larger parachute out of stowage 3: a funnel-shaped device which is attached to the end of a long flexible hose suspended from a tanker airplane in flight and into which the probe of another airplane in flight is fitted so as to receive fuel from the tanker airplane.

Although Webster's definition of the term "sea anchor" is consistent with its current usage, that for a drogue (in maritime use) only muddies the waters by referring to it as a sea anchor. This is the Voss syndrome all over again.

There is a need for consistent definitions of these two drag devices that have received a bum rap in the past when drogues were used over the bow and had virtually no capability of bringing the bow of the boat into the wind. It is important to differentiate between them because they are physically quite different and they perform different functions.

Confusion would reign on a boat if you did not differentiate between a halyard and a sheet; or between a spring line and a breast line; or between a genoa and a staysail. Yet all of these parts of a boat have great similarity in appearances, but are used differently.

A *sea anchor* is a large drag device deployed over the *bow* of a boat to hold the bow into the wind and the wave. It has to be large enough to prevent the bow from falling off and allowing the boat to lie broadside to the waves. A boat riding to a sea anchor will experience some leeward drift depending on the relative size of the sea anchor and the boat and the fury of the storm. A proper sea anchor will hold the backward drift to under 1 or 2 knots maximum. To allow the boat to drift aft at a greater speed is to put the rudder at risk. A sea anchor is used when a boat is disabled or the crew no longer wishes to sail, but would simply like to hold a relatively safe position and attitude with respect to the seas.

A *drogue* is a smaller drag-producing device that is deployed over the *stern* of a boat to slow its forward progress when running downwind and, to some extent, to hold the stern to the seas. In general, the drag of a drogue is considerably less than that of a sea anchor; hence, the forward speed of a boat towing a drogue will be considerably greater than the backward speed of a sea-anchored boat. When using a drogue, rudder control must be maintained so a compatible boat speed must be preserved. This may be as high as 5 or 10 knots but probably never less than 3 knots. A drogue is used when the crew feels that it is necessary to slow the forward progress of the boat while running downwind without surrendering control to the elements and while there is plenty of sea room to safely continue progress.

The differences in deployment of these two drag devices are illustrated in Figure 3.

William Van Dorn, in his book *Oceanography and Seamanship,* says, "Of all common safety items, sea anchors (or drogues) have probably received the least engineering attention, as is well borne out by the variability of reports from those who have tried using them." But this situation is rapidly changing. Shewmon, Inc., of Safety Harbor, Florida, has been engaged for a number of years in developing and testing sea anchors and drogues. Engineering design products have been subjected to extensive tugboat pull-testing before being put on the market. Para-Anchors International of Summerland, California, has transferred parachute technology to sea anchors as well as created new designs. Blue Harbor, Inc., of Hopkins, Minnesota, has developed an interesting line of "Boat Brakes" for smaller boats. Hathaway, Raymond & Reiser of Stamford, Connecticut, has created an exceptionally novel drogue that was given the acid test in a Gulf Stream storm and came out with flying colors.

Of even more importance, though, is the research work being conducted by Donald Jordan under the sponsorship of the U. S. Coast Guard into the nature of breaking waves and their effect on boats and how drag devices can add to the boat's safety.

What has prompted this long overdue development activity in boat drag devices? In general it is due to the large increase in the number of boaters that are exposing themselves to Nature's fury while fishing, sailing, racing, and cruising. The need for this type of equipment has grown and it is now recognized as a viable product for the marketplace.

Even more than the general growth of boating was the impact of the 1979 Fastnet Race disaster when 24 boats either sank or were abandoned and 15 lives lost when a severe gale passed through the fleet. It is fantasy to conjecture whether any of those lives could have been saved if the boats had been equipped with sea anchors and/or drogues. We do know, though, that one boat, *Windswept,* did deploy a drogue and survived the event in spite of serious damage and crew injuries. Most of the reasons given later for crews having to abandon boats—dismastings, loss of rudders, capsizing, and intolerable motions belowdecks—are conditions which often can be alleviated through the use of either a sea anchor or a drogue.

The use of a drag device should be considered when winds and seas cause the helmsman's knuckles to turn white. Common sense at some point must overrule macho determination to complete a race or a passage. Van Dorn offers two quantitative measures of when to consider your vessel in danger from the seas: The first is when the wave height nears a match with the width of the vessel's beam. At that point it is dangerous to take the seas beam-on. To continue sailing across the wind is to invite a capsize or, even worse, a 360° roll. The second measure applies when the wave height nears a match with the vessel's length. At that point a pitch-poling of the boat becomes a distinct possibility.

It is critical to the well-being of the boat and crew for the skipper to make tactical decisions to ward off potential disaster when the wave height reaches the width of the boat's beam. Possible tactics would involve the use of a sea anchor or a drogue to keep the boat aligned with waves either bow-on or stern-on with speed moderated.

Heaving-to or lying ahull in waves of lesser height than one beam width has been a successful tactic for many years. Heaving-to or lying ahull when the waves reach or exceed one beam width in height would seem to be courting disaster. If there can be any argument here, it is only in estimating wave height under adverse circumstances. For those who

A 6-foot-diameter Mark III Shewmon sea anchor on the aft deck of a harbor tug used in the testing program. Courtesy Shewmon, Inc.

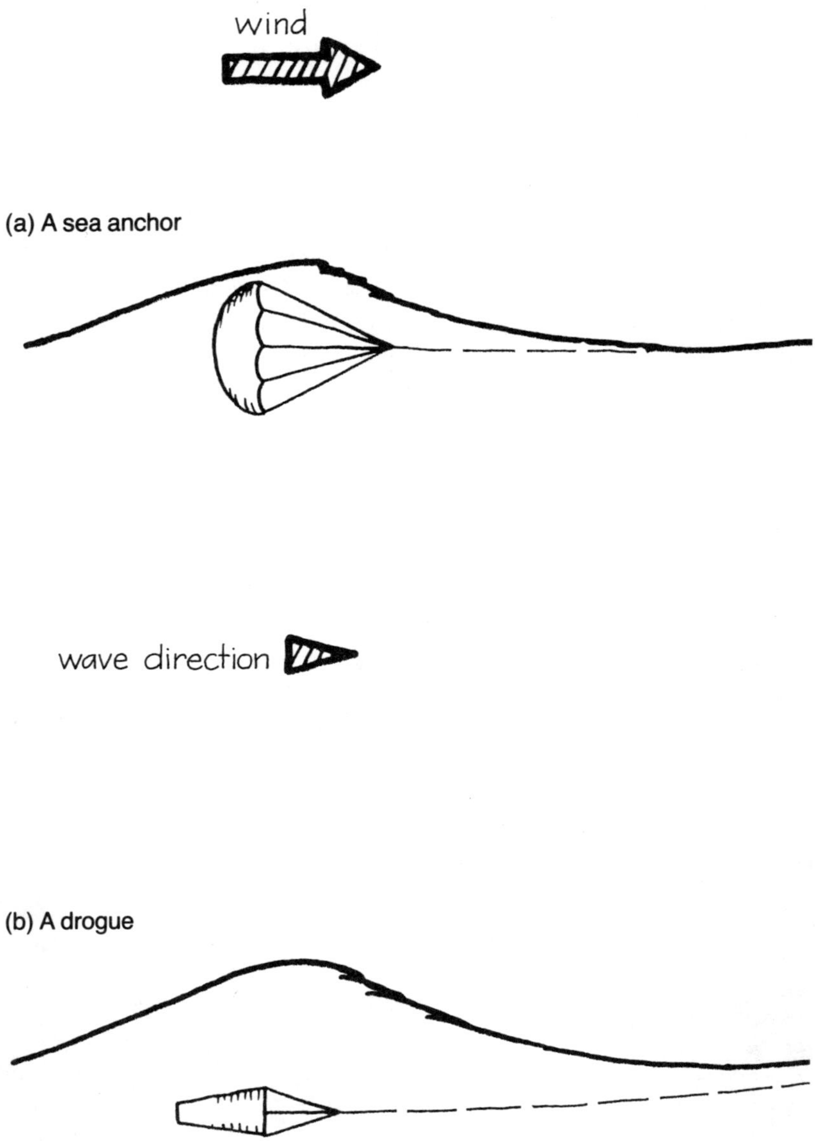

Figure 3. Using drag devices in a storm.

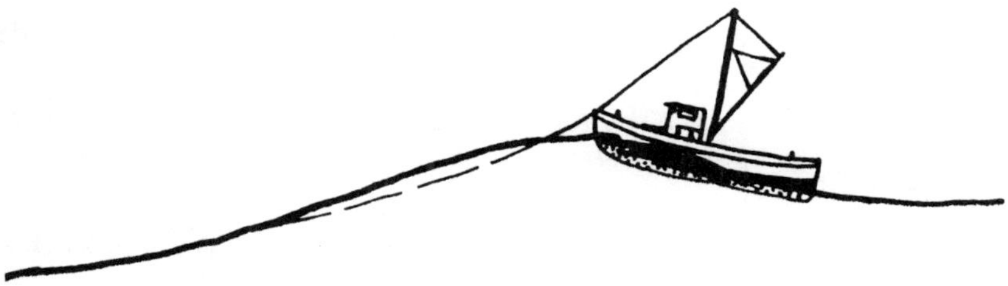

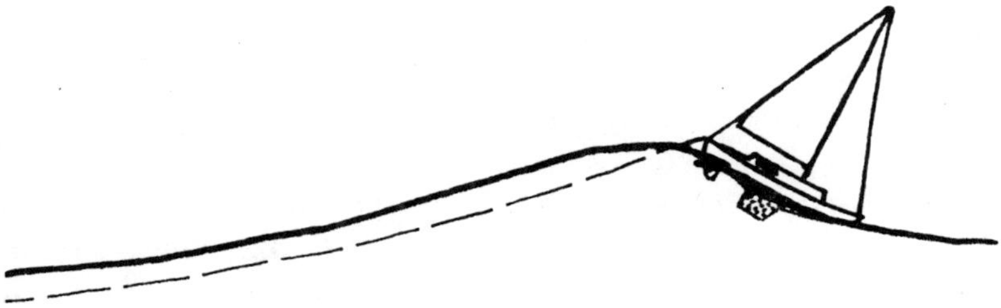

may think their vessels are impervious to Van Dorn's rule of thumb, tank tests (Jordan, *Sail* magazine, June 1986, September 1984, December 1982, and February 1982) and numerous actual experiences (e.g., the 1979 Fastnet Race) seem to verify the rule.

In the 1979 Fastnet Race it was the smaller boats which were capsized or pitchpoled, indicating at least a qualitative relationship between boat size and wave height. Since the waves were so large and confused, there was no way to relate the boat's beam or length to wave height. Conditions were well beyond that simple relationship. One boat in the 44- to 55-foot class had to be abandoned, but otherwise the casualties occurred in the classes 38 feet in length and less (Rousmaniere 1980).

One use for the sea anchor (not the drogue) for which there is little argument is the situation where the vessel is faced with being blown onto a lee shore. While there remains a safe sea distance, a large sea anchor set to windward will give the boat a new lease on life and the crew an opportunity to solve the problem of how to escape the lee shore. The few hundred dollars that the skipper invested in a proper sea anchor could be the sweetest insurance he ever purchased.

Probably the most common reason for carrying a sea anchor would be to stabilize the boat in case of a major failure short of sinking. In the event that a powerboat loses its primary propulsion, a deployed sea anchor (not a drogue) will keep its bow to the waves and permit repairs to be made to the power plant. There is no comparison to the ease of working inside a boat when it is bow to the waves rather than rolling in a trough.

Maximum comfort and a measure of safety in heavy seas can only be found on a boat when the bow is pointed into the waves and the roll is reduced to a minimum. That is the domain of the sea anchor.

For the blue water sailor who is racing or cruising, the value of sea anchors and drogues in common mishaps should not be underestimated. Take the case of a dismasting, or blown sails, or a broken rudder. To lie helpless without propulsion or steerage is a fate worse than death—an unwelcome, very possible alternative if the 1979 Fastnet Race is any indication.

Take, for instance, the saga of *Ipo,* a 30-foot sloop out of Honolulu that departed Lahaina, Maui, for her home port in forecasted mild weather. Soon after departure, the weather changed for the worse. A whole gale descended from the north catching the boat with all sails standing and tearing mainsail and jib beyond repair. With no other sails on board, the skipper of *Ipo* decided to motor back into port, only to find that the iron genny was also under the weather. Now there was nothing to do but radio for help, for the north wind was blowing *Ipo* away from the islands and the next stop would have been Palmyra atoll, 1,000 miles south. The Coast Guard, as they do so often, saved the day and brought *Ipo* into port.

It is conceivable that a sea anchor could have helped in *Ipo*'s situation. Having it hold the bow to the seas when the engine failed may have allowed the crew to make essential repairs to get the engine started. Certainly a sea anchor would have reduced the drift of the boat, which was over 60 miles in 24 hours. While double propulsion failures (sails and engine) are rare, nevertheless, the sea anchor makes sense for self-sufficiency.

The drogue comes into its own on still-functioning boats when it is desired to continue sailing or powering downwind and downwave. A drogue set over the stern can moderate the speed to a controllable level as well as help hold the stern into the waves to prevent broaching.

Sailors racing modern light displacement sailboats find that in gale conditions it is often better to head downwind and keep the boat sailing rather than fighting upwind, lying ahull, or heaving-to. The problem then becomes one of keeping the boat's speed under control, as light displacement boats pick up on the waves and like to go surfing. As a boat's speed approaches wave speed (10 to 20 knots) there is a great danger of broaching or pitchpoling. Either spells disaster. But there is a solution—a drogue over the stern to moderate the boat's speed without stopping it. A proper drogue will still allow the boat to sail fast downwind but it will be under better control and will give the helmsman more time to make corrections as the overtaking seas change shape and direction.

Thomas Seabrook in critiquing survival tactics used in the Fastnet Race said (*Yacht Racing/Cruising,* February 1983):

> Boats whose skippers chose to *lie ahull* under bare poles and therefore were unable to take action to avoid particularly dangerous seas were worse off than those that were kept under speed and steering control. The strong consensus among skippers was that the helm should be manned at all times. From this it would seem active rather than passive tactics will be more successful during a severe storm.

In another application a drogue can be very useful when running river mouths or inlets to harbors where sand bars create short, steep waves. If you absolutely must run a bar that is breaking, your chances of successfully doing so are immeasurably improved if a proper drogue is deployed before the entrance is reached. Once you enter the waters of the bar, your fate is in the hands of the helmsman, the drogue, and the Almighty.

Like all anchoring skills, practice is essential to the proper deployment of a drag device, either a sea anchor or a drogue. The environment is at its worst when the need arises, and that is no time to learn correct deployment procedures. If you have practiced your sea anchor and drogue deployment under less pressing circumstances, the chances are that you can make a proper deployment when the chips are down.

After an extensive review of the literature on sea anchors, drogues, and storm survival, and after having talked with many persons on the subject, I believe that bad experiences with these devices have been the result of improper use. On the other hand, the proper use of a drag device has time and again shown itself as a way to storm survival.

It is not within the scope of this book to get into the broader subject of tactics to be used during heavy weather. The reader is referred to K. Adlard Coles's book, *Heavy Weather Sailing,* for discussions on heaving-to, lying ahull, running at high speeds, and other survival techniques applicable to medium to heavy displacement boats. For light displacement boats the reader is referred to John Rousmaniere's book, *Fastnet, Force 10,* which describes in vivid detail the techniques used by several racing sailboats to survive what was the most disastrous racing event in the history of the sport.

Shed all sailors' tales when you read the present book and think objectively about the differences between sea anchors and drogues. Then you will be prepared to choose the drag device that offers the best chance for survival when you encounter the ultimate storm.

TWO

Deep Water Wave Characteristics

Waves in deep water are generated by the wind and their shape is unaffected by the proximity of the bottom. The general shape of a wave at low wind speeds is that of a mathematical sine wave, Figure 4. As the wind increases in strength, the shape changes to a trochoid and then to a cycloid at speeds approaching 60 knots. Over 60 knots the seas become so confused that simple mathematical representations are no longer valid. Let us take a look at the sinusoidal wave form in order to develop our language of waves.

We see in Figure 5 a simple wave shape with representations of wave length L, wave height H, and depth of water D. The crests are the "hills" of the wave and the troughs are the "valleys" of the wave. The water is sufficiently deep that there is no influence of the bottom on the wave shape.

Wave heights are very difficult to measure in the real sea. Bowditch in the *American Practical Navigator* has developed a wave height scale that corresponds with the wind speed on the Beaufort scale. This is reproduced in Table 1. The storm conditions we are particularly interested in are those for Beaufort Number 7 and above. For those winds we see that wave heights start at 13 feet and grow to phenomenal heights of over 45 feet. Few of us will ever see a hurricane at sea, but Beaufort Force 11 storms are not uncommon for the blue water cruiser or the long distance ocean racer.

Besides the height of a wave, the motion of the water particles at or near the surface is very important to the use of sea anchors or drogues. We have been told that the water near the surface of a wave does not move forward with the wave, only up and down. That is sufficiently correct for a general understanding of waves but not sufficiently exact for designing a sea anchor system for survival. Actually, particles of water within a wave rotate in an orbit at any location, going up and down and

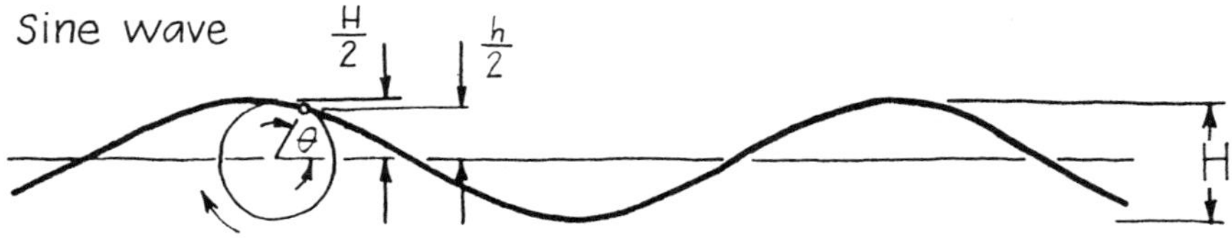

The sine wave is generated by plotting the vertical height above and below a horizontal line of a point on the rim of a rotating wheel. As the wheel rotates, the height varies as the trigonometic function: $\frac{h}{2} = \frac{H}{2} \sin \theta$.

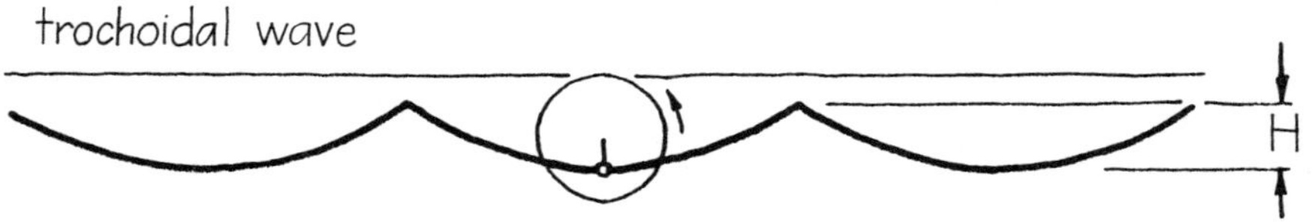

The trochoidal wave is generated by a wheel rolling along the underside of a line, with a fixed point on one spoke tracing the wave shape.

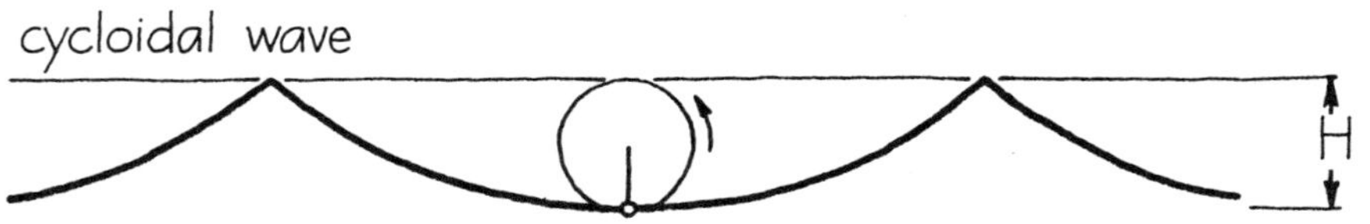

The cycloidal wave is created in a manner similar to the trochoidal wave, except that the point generating the wave is at the rim of the wheel.

Figure 4. Mathematical representations of deep water wave shapes.

forward and backward as the wave passes by. This orbital motion is illustrated in Figure 6.

The diameter d of the orbit is equal to the height H of the wave, and the speed is such that a complete revolution of a particle takes place in a time equal to the period P of the wave. The period of the wave is the time it takes one full wave (from crest to crest) to pass a particular point in space. This orbital motion of water particles extends downward as far as one wave length in depth, but for all practical purposes it has vanished at a depth equal to one-half a wave length. From this relationship we can define deep water as water depth in excess of one-half a wave length for the particular fully developed wave system of interest. If the water is of

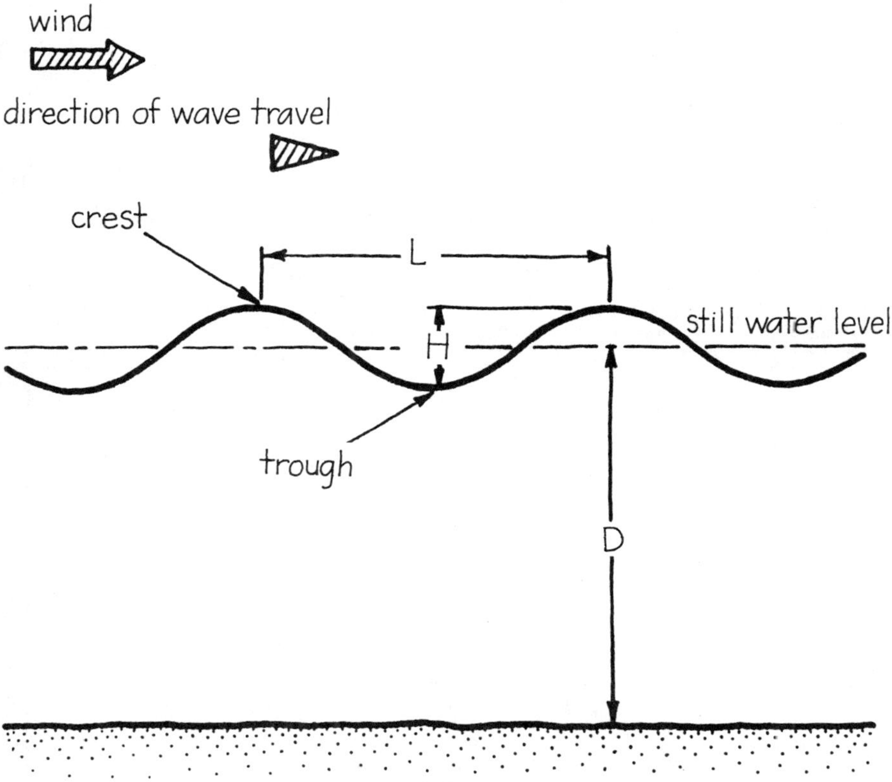

Figure 5. Deep water wave characteristics.

lesser depth, then bottom effects become noticeable, making the waves steeper and closer together. These types of waves are more appropriate for surfboards than for boats.

There are some basic, deep water wave characteristics which are of interest to us for designing a sea anchor or drogue system. These are the period, length, speed, and height of the wave as well as the orbital speed of the water particles at the wave surface. The following mathematical relationships describe the wave motion of interest to us:

Length of a wave: $L = 5.12\ P^2$ or $\dfrac{S^2}{1.8}$ [*Equation 1*]

Speed of a wave: $S = 1.36\ L^{1/2}$ [*Equation 2*]
or $S = 3.03\ P$

Orbital speed of a particle: $\Delta s = \dfrac{1.9\ H}{P}$ [*Equation 3*]

where L = wave length (feet)
P = wave period (seconds)
S = wave speed (knots)
Δs = particle orbital speed (knots)
H = wave height (feet)

Numerical values of the above deep water wave relationships can be extracted from tables in Bowditch or other oceanographic references.

Beaufort number	Wind speed knots	Seaman's term	Estimating wind speed Effects observed at sea	Term and height of wave - ft. *	Sea state code
0	under 1	Calm	Sea like mirror.	Calm, glassy, 0	0
1	1-3	Light air	Ripples with appearance of scales; no foam crests.		
2	4-6	Light breeze	Small wavelets; crests of glassy appearance, not breaking.	Rippled, 0-1	1
3	7-10	Gentle breeze	Large wavelets; crests begin to break; scattered whitecaps.	Wavelets, 1-2	2
4	11-16	Moderate breeze	Small waves, becoming longer; numerous whitecaps.	Slight, 2-4	3
5	17-21	Fresh breeze	Moderate waves, taking longer form; many whitecaps; some spray.	Moderate, 4-8	4
6	22-27	Strong breeze	Larger waves forming; whitecaps everywhere; more spray.	Rough, 8-13	5
7	28-33	Moderate gale	Sea heaps up; white foam from breaking waves begins to be blown in streaks.	Very rough, 13-20	6
8	34-40	Fresh gale	Moderately high waves of greater length; edges of crests begin to break into spindrift; foam is blown in well-marked streaks.		
9	41-47	Strong gale	High waves; sea begins to roll; dense streaks of foam; spray may reduce visibility.		
10	48-55	Whole gale	Very high waves with overhanging crests; sea takes white appearance as foam is blown in very dense streaks; rolling is heavy and visibility reduced.	High, 20-30	7
11	56-63	Storm	Exceptionally high waves; sea covered with white foam patches; visibility still more reduced.	Very high 30-45	8
12	64-71	Hurricane	Air filled with foam; sea completely white with driving spray; visibility greatly reduced.	Phenomenal, over 45	9

Table 1. Wave heights by the Beaufort scale, from Bowditch.

*For a fully developed sea state.

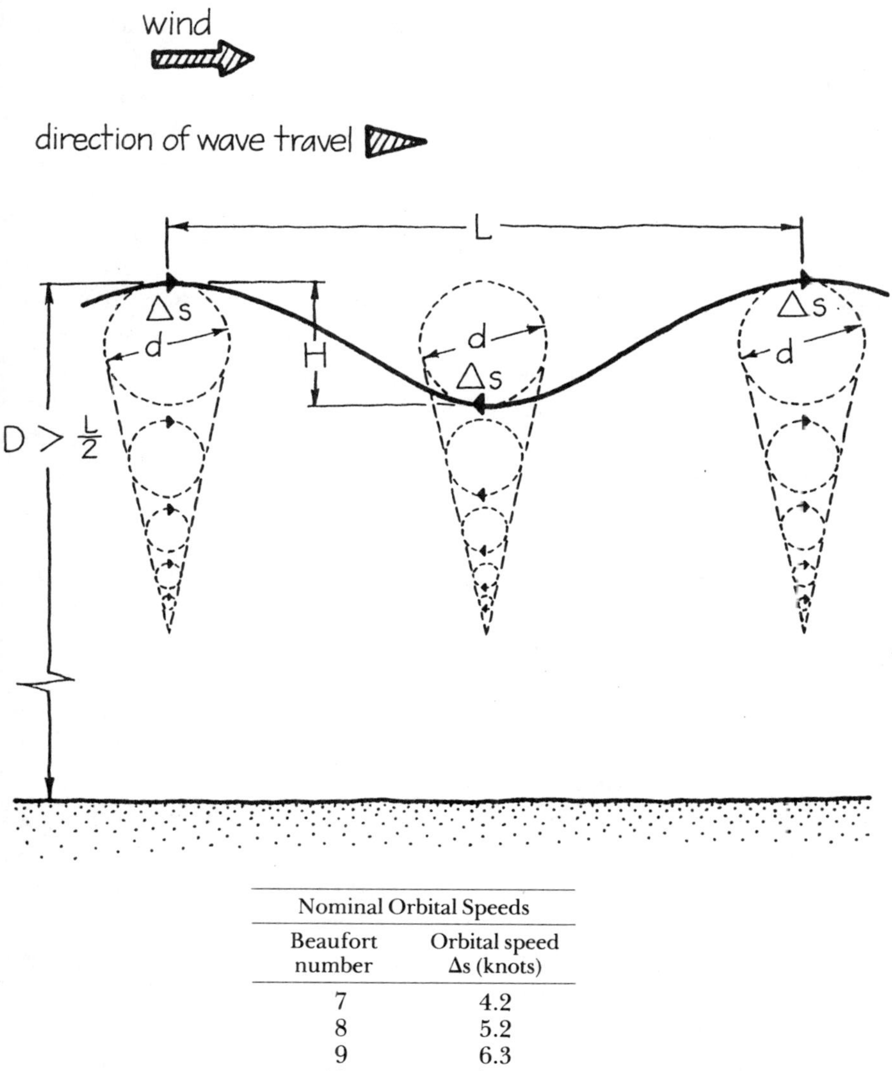

Beaufort number	Orbital speed Δs (knots)
7	4.2
8	5.2
9	6.3
10	7.4
11	8.2

Figure 6. **Orbital movement of water particles in a deep water wave.**

As cycloidal waves become steeper, due to either an increasing wind or a shoaling bottom, the crests become sharper. The maximum theoretical sharpness of a crest before breaking is about 30° to the horizontal as illustrated in Figure 7 (a). As the wind further increases and the wave grows taller, the crest will fall over because the simple circular orbiting motion of water particles cannot keep up with the increase in the speed of the surface particles. The wave will then break in a very forceful manner.

Two types of break are possible: the sliding break and the curling break. The sliding break shown in Figure 7 (b) is typical of the open ocean of great depth where there is usually confusion of waves. The overfall can be very turbulent and forceful. The curling break of Figure 7 (c) is more typical of shoaling water and is the delight of surfers who seek to ride the "pipeline." The weight of water in the curl can be awesome and, after it breaks, it will degenerate into the sliding type of break.

The speed of the breaking crest, Δs_b, in Figure 7 (d), is the speed of advance of the wave itself. This can be visualized as the transfer of the orbital momentum of the non-breaking wave of equal speed into linear momentum at the crest of the breaking wave. The magnitude of the breaking speed, which is what the boat would feel, is tabulated in Table 2, for the case of a 60-knot (Force 11) wind in deep water.

Van Dorn gives the following rules of thumb for the frequency of breaking waves: For wind speeds under 10 knots, less than 1 percent of the waves break. For wind speeds over 60 knots, 100 percent of the waves break. Between these two extremes there is a linear variation of breaking probability.

The power unleashed by a breaking wave should not be underestimated, nor should it be cause for panic. When waves begin to break, it is time to weigh the capabilities of the crew and the boat to continue the present passage or to take action by setting a sea anchor or a drogue.

You should not delude yourself into thinking that the ocean surface is always made up of well-defined and well-behaved wave patterns. It is not. At low wind speeds, you will be able to identify wave height, length, and period. But as the wind increases in strength, confusion grows until it reaches the chaotic maelstrom of hurricane-driven waves. Then you will be thankful that you paid earlier attention to alternative means for surviving such storms.

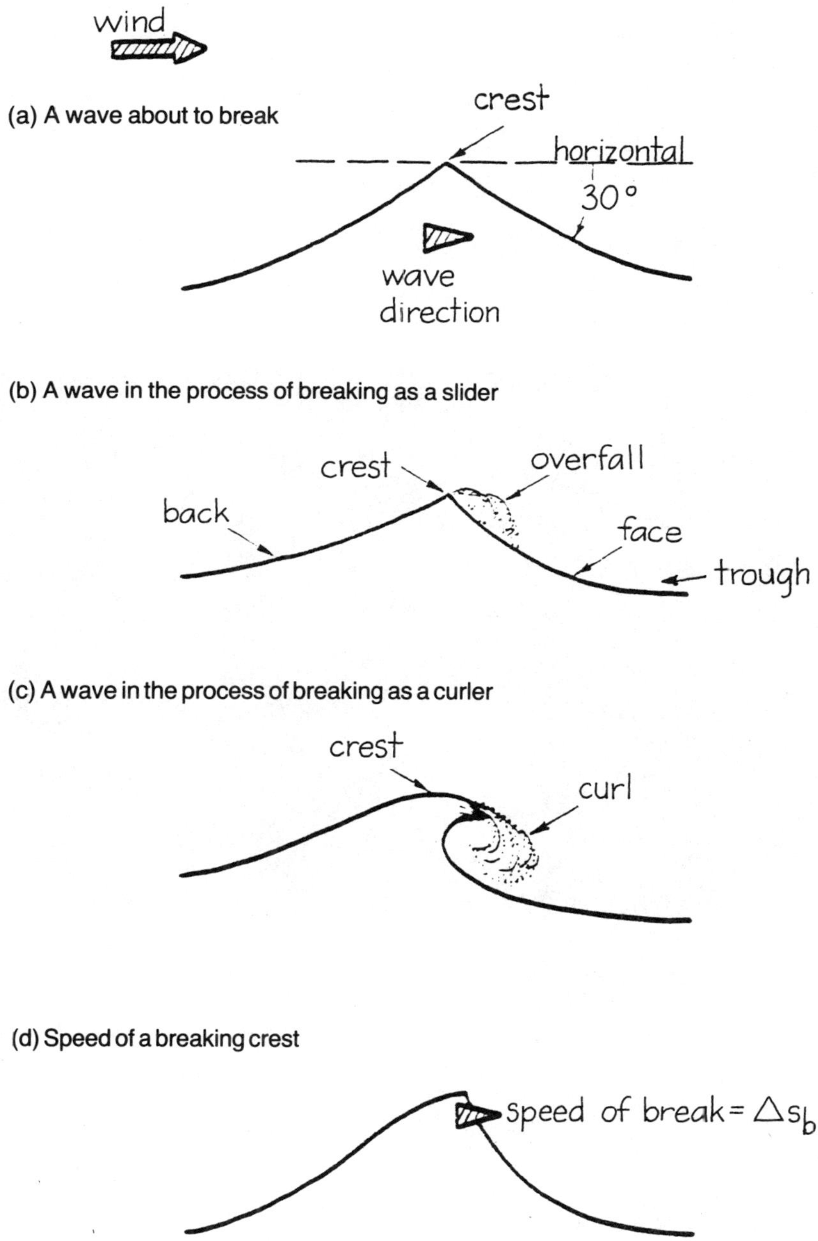

Figure 7. Characteristics of breaking waves.

Duration of blow T (hours)	Wave length L (feet) in deep water	Wave height H (feet)	Speed of break Δs_b (knots)
1½	100	9	13
4	200	19	18
8	300	30	23
11½	400	38	27

Table 2. Approximate characteristics of breaking waves for 60-knot winds (Force 11).

Under the right circumstances a breaking wave can be a beautiful work of Nature. Courtesy *Latitude 38*.

THREE

Sea Anchor Design

Sea anchors are large drag-producing devices that are deployed over the bow of the boat to hold the bow into the wind and minimize the backward movement of the boat. It is necessary that the sea anchor be a relatively large device in order to "anchor" the boat to the surface water which is obviously fluid and yielding. (Whereas a drogue is intended to only slow a boat down in its normal forward motion, a sea anchor is expected to essentially hold it from moving on the sea's surface.) A very large parachute-type sea anchor can do this task, but since the water itself is in motion due to wind-driven currents, the boat will still move over the bottom at the speed of the local current.

A boat riding to a large sea anchor has a motion very similar to that of a boat riding to a ground anchor. The biggest difference is in the nature of the wave forms themselves. At sea the waves tend to be longer and not as steep as the waves in shoaling waters but they will still crest frequently. The powerful but irregular motion of the sea when resisted by a sea anchor can produce great strains on the sea anchor gear. The dynamic loading will give all of the problems found in ground anchoring such as overloading cleats and Samson posts, causing severe chafe on the tether, and occasionally burying the bow in green water. The one mitigating factor of the sea anchor streamed from the bow is that it keeps the bow pointed into the wind which is the normal riding position for the boat.

Cruising literature abounds with instances of monohull boats riding out severe storms by lying ahull and letting the boat seek its own comfortable riding position with respect to wind and wave. This apparently is no longer true since the modern boat design, with its center of underwater lateral resistance well aft and a great deal of side windage forward, tends not to weathercock into the wind. Instead, it takes up a beam-on position, life belowdecks becomes untenable, and the boat is liable to capsize if hit by a breaking sea.

Multihull sailors were the first to make a concerted effort to improve the safety of their craft through the use of sea anchors. They adapted military surplus parachutes as sea anchors with significant success. Randy Thomas writes in *Yachting* (November 1985) of his circumnavigation:

> On the afternoon of July 25, 1985, *Celerity* sailed out of the fog in Juan de Fuca Strait and secured to the same Victoria wharf from which she had departed on her Pacific cruise eight years before. To complete her Pacific circumnavigation, the 31 foot trimaran became the first multihull to make the non-stop high-latitude crossing from Japan to mainland North America.
>
> Hard on the wind for 17 days, *Celerity* also encountered 800 miles of calms and rode out four gales on her parachute sea anchor. She logged approximately 4500 miles, completing the crossing in 48 days without the use of auxiliary power.

The airman's parachute adapted for anchoring at sea. Note the many shrouds and the small vent at the crown of the canopy. This 24-foot-diameter sea anchor stows in less than 1 cubic foot of space. Courtesy Para-Tech Engineering Co.

There is nothing in boat design that prevents powerboats or sailing monohulls from also using sea anchors when the need arises. A properly sized sea anchor is essentially a parking brake and as such can be used in a variety of situations. Commercial fishermen use them often to hold their boats near the fishing grounds during the night while they sleep. Powerboats of all descriptions have used them to hold their bows into the wind and minimize drift while they repair propulsion machinery so that they can get underway again. The difference in working on a cranky engine in a boat when it is held bow to the wind, as opposed to free-drifting sideways, can be the difference between getting the engine back in operation quickly or succumbing to mal de mer.

Powerboats experiencing a loss of power near a leeward shore can be in a most difficult situation and must buy time to either get the engine running again or seek help from outside sources. While a sea anchor will not cut the boat's drift to zero, it can provide the skipper with additional time to solve his problem. Certainly any single-engine powerboat would benefit in an emergency by having a sea anchor aboard.

For any boat a sea anchor bringing the bow of the boat into the wind will make a more pleasant ride whether fishing, sunning, sleeping, or doing maintenance. The applicability of sea anchors to work- and recreational boats is not yet fully developed, especially in the case of monohull sailboats, but their seeming ability to save boats under adverse situations is a powerful incentive for more development on all fronts.

Most of the condemnation of sea anchors mentioned in boating literature has been a result of too small a sea anchor having been used. A sea anchor must be large enough to prevent the bow of the boat from being driven off to leeward by wind and wave. A riding sail is often used to help prevent the boat from yawing.*

Some writers say that they want a sea anchor with "give" to it—they don't want to be held firmly against the impact of a wave. The idea that a properly sized sea anchor will "give" with the waves is patently untrue. The large mass of water in a sea anchor† cannot be accelerated rapidly enough to absorb a wave's impact on the hull. A small sea anchor only lets the boat drift to leeward faster and to yaw into a broaching position. As

---

*Yawing is the angular motion of a boat about its vertical axis. When the boat yaws, the bow and stern swing from side to side.

†The mass of water in a 9-foot-diameter hemispherically shaped container is approximately 190 cubic feet, or 12,000 pounds of seawater.

the sea anchor becomes smaller and smaller, the boat begins reacting more and more like a boat that is lying ahull.

Did you ever think that a larger ground anchor would transmit increased loads to your boat? Certainly not; all it does is assure that you have a more positive grip on the seabed. Similarly, a larger sea anchor gets a better "bite" on the water and is in a better position to withstand the constant loads of the wind on the boat. The strains on tether and hull will be no greater than with a small sea anchor that is not slipping. In fact, a small sea anchor that lets the boat yaw excessively will cause a greater strain on the boat when a wave hits it broadside at its greater angle of yaw.

What, then, is different about using a large sea anchor in place of a ground anchor? First, most ground anchoring takes place in the lee of some land, or at least where there is a reef or breakwater to moderate the wave action. Waves, and not wind, are the major factor in either type of anchoring. Where a boat is ground anchored without the benefit of shelter from the seas proportionate to hull size, the waves will be steeper and higher than in the open ocean because of the shoaling waters. That is why boats seek anchorages sheltered from the seas with only secondary regard for shelter from the wind.

In the open sea there is no relief from the waves but they have a longer period and tend to break less frequently than shoal water waves. Sea anchoring with adequate sea room has to be a less worrisome problem than ground anchoring with the shoreline only feet away.

TYPES OF SEA ANCHORS

In essence, all sea anchors look something like an airman's parachute when in use and, in fact, some are parachutes. A sea anchor has also been referred to as a hydraulic parachute. The independent variable of the sea anchor is its size and, hence, its drag. The dependent variable is its strength and there is no question that a sea anchor has to be strong and durable. Worse than having no sea anchor at all is having one that fails in the midst of a storm, allowing the boat to broach and present a vulnerable side to the onslaught of breaking waves.

Parachute Sea Anchors

The first true sea anchors were airmen's parachutes. (I do not consider the Voss or Fenger types of drogues to be sea anchors, regardless of what their inventors called them.) By itself the parachute is a drag device meant

The 10-foot-diameter Shewmon Mark III sea anchor was designed for use with ships as large as destroyers and has a strength of 60,000 pounds. The latest Mark-9 design has been made as large as 27 feet in diameter. Courtesy Shewmon, Inc.

to slow the descent of a stricken aviator to some moderate speed that will enable a safe impact. The same parachute when deployed in water develops a drag 800 times as high due to the greater density of seawater. That drag is sufficiently large to counteract the wind loads on most recreational size vessels. Whereas an airman's parachute under the force of gravity allows a descent speed anywhere from 15 to 20 feet per second, the same parachute operating in water will slow the vessel's drift to 1 or 2 knots at most.

The parachute adapted as a sea anchor does away with the body harness and risers but retains the shrouds and canopy intact. The typical parachute canopy is made of rip-stop nylon (much like spinnaker material) and is made in many gores to give it a hemispherical shape. Shroud lines and gores are sewn together with the shroud lines passing completely over the top to give the necessary strength. An 18-inch-diameter vent hole is left in the top of the canopy to let some air escape and give stability to the parachute. Reinforcing patches (these may be made of Kevlar) are strategically located at points of load concentration.

Parachutes now used as sea anchors are all of the hemispherical type. Recently, highly maneuverable rectangular parachutes have come on the scene, but their use as sea anchors has yet to be tested. The standard hemispherical parachute when new costs upwards of $1,000, but used and surplus parachutes can be purchased for as low as $50. The saving is most attractive and the parachutes are still in good shape. Parachutes professionally converted to sea anchors by knowledgeable companies retail for $150 to $200.

Useful rules of thumb have been developed over the years for the proper size parachute to use as a sea anchor. The basic airman's parachute comes in a range of sizes—20, 24, 26, 28, 30, 32, 34, and 36-foot diameters. All of these, however, may not be available at the surplus store at any one time.

It was John and Joan Casanova who developed the first rule of thumb for the size of a parachute to sea anchor a multihull boat. They determined that the diameter must be equal to or surpass the boat's beam (Casanova et al. 1982). Their *Tortuga Too* used a 28-foot-diameter parachute sea anchor as standard equipment and survived 80-knot winds off New Zealand and 40-foot seas in the Indian Ocean. They related that the weather was so bad in the Indian Ocean that they spent 1 out of every 6 days tethered to their 28-foot-diameter parachute sea anchor.

Para-Anchors International recommends the following sizes of parachute sea anchors for multihulls and monohulls:

24-foot-diameter canopy:	Multihulls to 24-foot beam and monohulls to 45 feet LOA
28-foot-diameter canopy:	Multihulls to 28-foot beam and monohulls to 55 feet LOA

Experience has shown that multihull boats lie to a sea anchor better than do monohull boats. This is most likely due to the wide-based bridle attached to the outriding hulls. The narrow bow of the monohull precludes as effective a bridle spread as in the multihulls but, nevertheless, ground anchoring experience has shown that even a narrow bridle spread reduces the amount of sheering that takes place.

Two events reported by Para-Anchors International occurred in 1984 that are worthy of note. They supplied a J/30 sailboat with a 9-foot Para-Anchor and, in an ensuing storm, the Para-Anchor pulled the J's bow into 20-foot seas and kept it there for 4 hours during which time the crew members were able to rest.

In the other instance a 76-foot steel ketch of 36 tons displacement with a full keel rode to a 28-foot parachute sea anchor for 18 hours in a gale off the Azores. Winds were 55 knots and seas to 20 feet. The skipper reported that "the Para-Anchor worked perfectly, we rode nicely. I will never make an ocean passage without one on board." (*Force 10 Imminent,* August 1985)

Parachute sea anchors are made from the common personnel parachute sizes used by airmen but even larger parachutes have been made for cargo drops from airplanes. A 35-foot-diameter cargo parachute is common, and Van Dorn describes the use of a 100-foot-diameter parachute as a "drogue" [sic] on a weapons test barge.

Shewmon Sea Anchors

The Shewmon company of Safety Harbor, Florida, manufactures a line of sea anchors and drogues designed and fabricated for those specific purposes. They are designed to insure self-opening and to produce maximum pull at minimum drift speeds. The canopy of the Shewmon sea anchor is made from 7½-ounce polyester cloth heavily reinforced at critical stress points. Canopy ropes are stitched into the canopy panel joints with Dacron thread, providing excellent carry-through strength from the panels to the shroud lines.

The strength, performance, and stability of the Shewmon sea anchors have been verified by the manufacturer's testing program in which developmental and production sea anchors have been towed behind powerful harbor tugs at speeds to 8½ knots. Besides confirming the strength and drag of the sea anchors, the tests have shown them to be reasonably stable in operation.

In a technical report the manufacturer gives an analytical method for estimating wind drag forces on a boat and water drag counterforces, which combine to set the pull of the sea anchor. Two numerical examples of drifting speeds are given for boats bow to windward:

36-foot sloop in 60-knot winds:	.94 knots with a 5-foot sea anchor .72 knots with a 6-foot sea anchor and a steady state pull of 680 pounds
48-foot motor yacht in 60-knot winds:	1.7 knots with a 5-foot sea anchor 1.4 knots with a 6-foot sea anchor and a steady state pull of 1,600 pounds

The above examples would indicate that a 5-foot sea anchor is suitable for a 36-foot sloop in the winds considered, but the 48-foot motor

The 4-foot-diameter Shewmon Mark III sea anchor. Courtesy Shewmon, Inc.

yacht probably should have at least a 7-foot sea anchor to ensure that motion astern is never large enough to cause rudder damage.

SEA ANCHOR PERFORMANCE

The common shape of sea anchors is a truncated hemisphere, which provides an efficient drag shape with no requirements for stiffeners and which, if properly designed, will open automatically when deployed. The Shewmon company has done extensive testing on such sea anchors and has developed drag data for a variety of sizes and drift speeds ("Calculating Sea Anchor Performance" by Daniel Shewmon, *Cruising World,* June 1984, and "Survival with Sea Anchors" by Daniel Shewmon, *Cruising World,* March 1984). Dan Shewmon has kindly consented to allow the reproduction of these data in my own format, as shown in Figure 8.

Although there are no theoretical limits to the maximum value of the allowable drift speed, the practical limitation is how much backward loading a rudder can withstand before failure. I would suggest that the upper drift speed limit never exceed 2 knots. For overall design purposes I would suggest using a 1-knot drift speed.

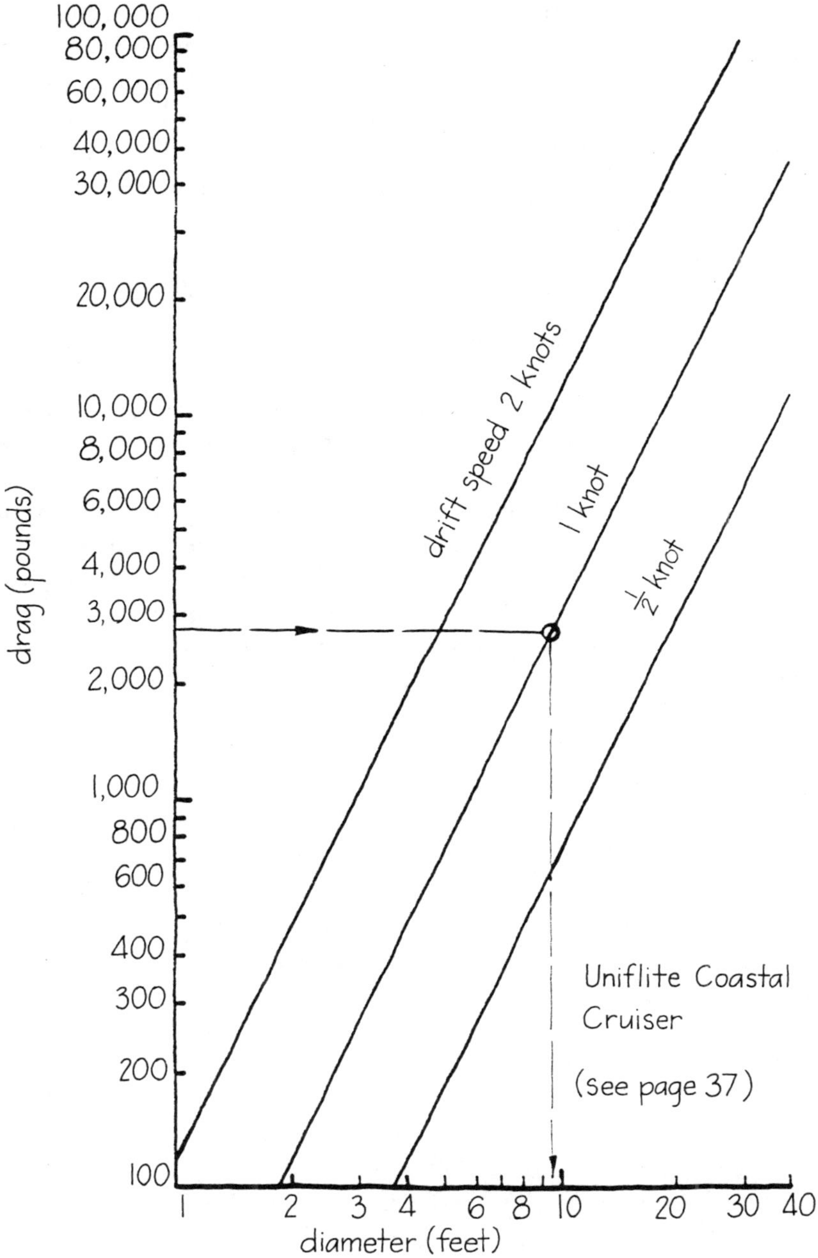

Figure 8. Sea anchor drag. By permission of Shewmon, Inc., © 1983.

Parachute-type drag devices are quite suitable as sea anchors where the speed through the water is low, say below a couple of knots. When higher speeds are encountered (as in drogue applications), the parachute-type drag device becomes directionally unstable, tending to move sideways and even to develop large amplitude lateral oscillations. This axial instability makes them generally unsuitable for drogue applications even in very small sizes.

ESTIMATING SEA ANCHOR SIZE

An analytical approach to sizing a sea anchor matches the wind drag of the boat to the water drag of the sea anchor. Since the sea anchor requires a little drift to develop holding power, it will slowly move to leeward, but at such a small relative rate as not to affect the wind drag above the waterline.

At the present time there is no closed analytical solution to determining sea anchor size, but there are several estimating techniques which can suffice until more knowledge is obtained. These sizing techniques are referred to here as the "wind loads method," "equal areas method," "average draft method," and "Van Dorn's method." All of them are approximations at best, but in the absence of more rigorous methods, they will have to do.

Wind Loads Method

One can design a sea anchor using the wind loads estimating technique developed in my book on anchoring (Hinz 1986). At sea anchor a boat will experience leeward drift since the sea anchor does not have a positive grip on the water such as the ground anchor has on terra firma. This leeward drift is predominantly due to the more or less steady pressure exerted by the wind on the hull, superstructure, and rigging. Non-breaking waves tend to produce a fore and aft oscillatory movement of the boat with little net change in position. The impact of the waves is absorbed by the elastic nature of the tether which, when it recovers with the boat on the backside of the wave, actually brings the boat near to its previous location.

If there is no leeward drift of the boat at sea anchor, then the sea anchor will simply float suspended in the water like a huge jellyfish. It has to have motion to develop a drag force and for design purposes you can safely assume 1 knot of leeward drift (V_d) in using Shewmon's data of Figure 8.

The aerodynamicist has given us various tools to estimate the wind loads on a boat at sea anchor. One method is to measure them in a wind tunnel, which few of us can afford to do. Another way is to estimate wind loads mathematically based on test data for other boats, for other vehicles such as automobiles and airplanes, and on a whole variety of components which make up a vehicle. The latter method will be described here as it can be done for most boat shapes using a simple hand calculator and the drawings of the boat. (A more complete discussion of determining the wind loads on an anchored boat is given in my book, *The Complete Book of Anchoring and Mooring* [Hinz 1986], from which this section is excerpted.)

The equation for the aerodynamic drag of an object in a wind stream is:

$$D_w = q \cdot C_d A \qquad \textit{[Equation 4]}$$

where D_w = drag force (pounds)
q = dynamic pressure (pounds per square foot)
C_d = drag coefficient (dimensionless)
A = characteristic area (square feet)

The dynamic pressure q is determined from the equation:

$$q = \frac{\rho}{2} \cdot V^2$$

where ρ = .002378 slugs per cubic foot at standard sea level conditions
V = wind speed in feet per second

Values of dynamic pressure q for a range of wind speeds at sea level are:

V (knots)	*Beaufort Number*	*q (pounds per square foot)*
10	3	0.33
20	5	1.3
30	7	3
40	8	5
50	10	8
60	11	12
70	12	16
80	13	21
90	15	27
100	16	33

The drag coefficient C_d comes from wind tunnel test data and is applicable to similar body shapes at different wind speeds as long as the flow is incompressible, which is the case for wind speeds of interest here (Table 3). These drag coefficients are for a 0° angle of attack and a 0° angle of yaw. The drag coefficients for boats include a nominal allowance for rigging, lifelines, and antennas.

The author calculated a number of additional drag coefficients for monohull sailboats of differing displacement-length ratios.

$$\frac{D}{L} = \frac{\text{displacement (long tons)}}{(.01\ \text{LWL})^3} \qquad [\textit{Equation 5}]$$

where LWL = waterline length (feet)

These correlated well with the displacement-length ratio of the hulls (Figure 9).

The characteristic area A is the area of the body measured perpendicular to the wind. The drag D of a body is proportional to the area of the body allowing scaling of drag forces for bodies of different size but the same shape. Boats, however, have been observed to veer as much as 30° from side to side when anchored. Therefore, the characteristic area to be used for wind drag load calculations is the yawed area presented to the

Vehicle	C_d with wind ahead
Angular tramp steamer	1.2
Cabin cruiser	1.0
Long-liner	0.9
Oil tanker	0.85
Modern ocean liner	0.7
Cruising trimaran	0.6
Cruising catamaran	0.55
Modern automobile (sedan)	0.52
Racing trimaran	0.45
Modern automobile (fastback)	0.34
Ultrastreamlined playboat	0.2
Racing automobile	0.17
Airplane	0.09

Table 3. Typical vehicle wind drag coefficients.*

*Source: T. Baumeister, ed. *Mark's Standard Handbook for Mechanical Engineers,* 8th ed. New York: McGraw-Hill Co., 1978; and William F. Durand, ed. *Aerodynamic Theory,* vol. 4. Gloucester, MA: Peter Smith, 1976.

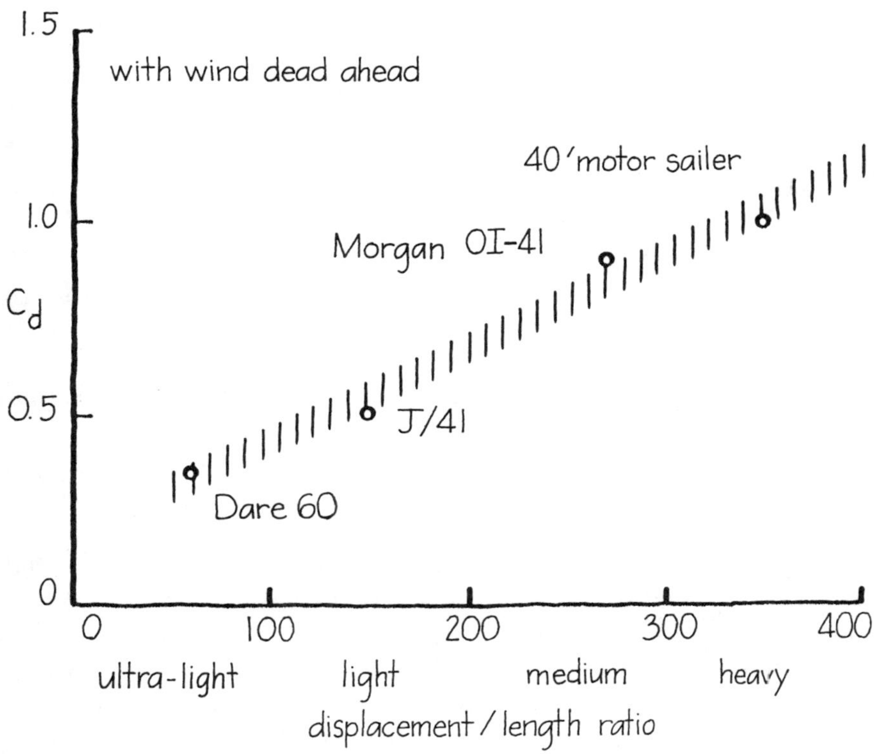

Figure 9. Monohull sailboat wind drag coefficients.

wind when the angle of yaw reaches 30°. One could also make a correction for the drag coefficient at an angle of attack but this becomes of secondary importance if the characteristic yawed area has been used.

The wind drag load on a boat yawed at 30° to the wind becomes:

$$D_w = q \cdot C_d \cdot A_{30°} \qquad \text{[Equation 6]}$$

Having determined the wind drag of the boat at a given wind speed, you then enter the graph of Figure 8 and determine the size of sea anchor needed to hold the leeward drift to your chosen value.

Example: The Uniflite Coastal Cruiser powerboat (Figure 10) has the following applicable dimensions:

Drag coefficient (C_d) = 1.0
Characteristic yawed area ($A_{30°}$) = 226 square feet

The drag force on this boat yawed 30° in 60 knots of wind is determined from Equation 6 as:

$$D_w = 12 \times 1.0 \times 226 = 2{,}700 \text{ pounds}$$

This is the amount of drag that the sea anchor must counter. For a designed leeward speed of 1 knot, Figure 8 shows us that the sea anchor must be 9½ feet in diameter.

If we designed for only 45 knots of wind, the drag would be 1,500 pounds and the sea anchor diameter would be 7 feet. Thus you can design your own sea anchor to match whatever you consider the highest wind speed that you will encounter. *Remember that a larger sea anchor will work at lower wind speeds but a smaller sea anchor may put your boat in jeopardy at higher wind speeds.*

If you have previously designed a ground anchor rode for your boat according to my anchoring book, then you can assume that the wind load D_w for the foregoing calculation is one-fourth of the breaking strength of the selected ground anchor rode and you do not have to repeat the wind load calculation of Equation 6. Since the boat is being faced with winds of the same magnitude in both ground and sea anchoring situations, the resulting wind loads are the same. (Note: The leeward speed of a boat at sea anchor should be less than 1 or 2 knots and is insignificant compared to the wind speed.) The factor of four in the strength of the anchor rode assures sufficient margin for the tether of the same size to take care of

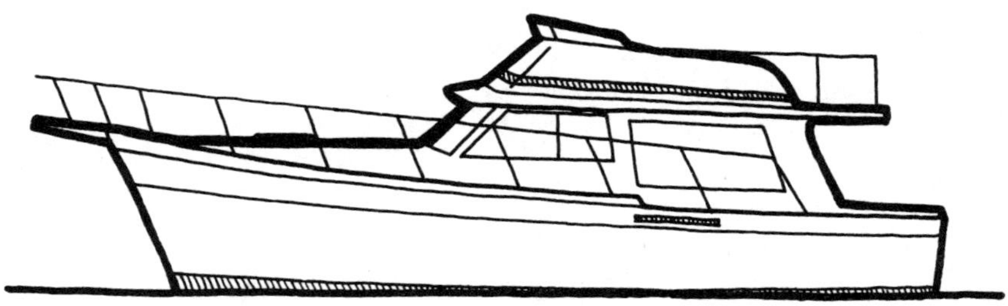

LOA 37 feet, 9 inches
Beam 12 feet, 9 inches
Draft 3 feet, 10 inches

Figure 10. Uniflite Coastal Cruiser.

wave impact and other variables. The length of the sea anchor tether, however, will probably differ from that used in your ground anchor rode. This will be covered later.

There is an advantage to calculating your sea anchor loads rather than guessing at them and it is that you also have the basis for determining the size of the tether and the associated connecting elements between the sea anchor and the boat.

Equal Areas Method

We have another hypothesis that says the sea anchor must be large enough to bring the bow of a boat lying broadside to the wind in a trough into the wind. This situation is illustrated in Figure 11. The turning torque produced by the sea anchor must be greater than the counter-torque produced by the wind drag acting forward of the center of lateral resistance. That is:

$$D_{sa} \times \ell_2 > D_w \times \ell_1$$

The greater this differential, the faster the boat will be turned bow-on to wind and wave. If we assume that the turbulent unit drag of the sea anchor and boat underbody are the same, then we can simply equate the area of the sea anchor to that of the underbody and come up with a first approximation of the required sea anchor size. The differences in lever arm lengths ℓ_1 and ℓ_2 assure that the sea anchor-induced torque will be greater than the wind load torque and that the bow will be pulled into the wind.

Once in the wind, the sea anchor will have more than enough drag to brake the leeward drift to 1 or 2 knots or less.

This equal area sea anchor sizing technique is easy to do, using a copy of the designer's profile view of the boat to determine the underbody silhouette area, and converting that to an equivalent circular area. The circular area of the sea anchor is converted to a diameter by the relation:

$$d_{sa} = \sqrt{\frac{4A_u}{\pi}} \qquad \text{[\textit{Equation 7}]}$$

where d_{sa} = diameter of the sea anchor (feet)
A_u = measured area of underwater profile of boat (square feet)

Example: The Out Island 41 medium displacement sailboat shown in Figure 11 has the following applicable dimension:

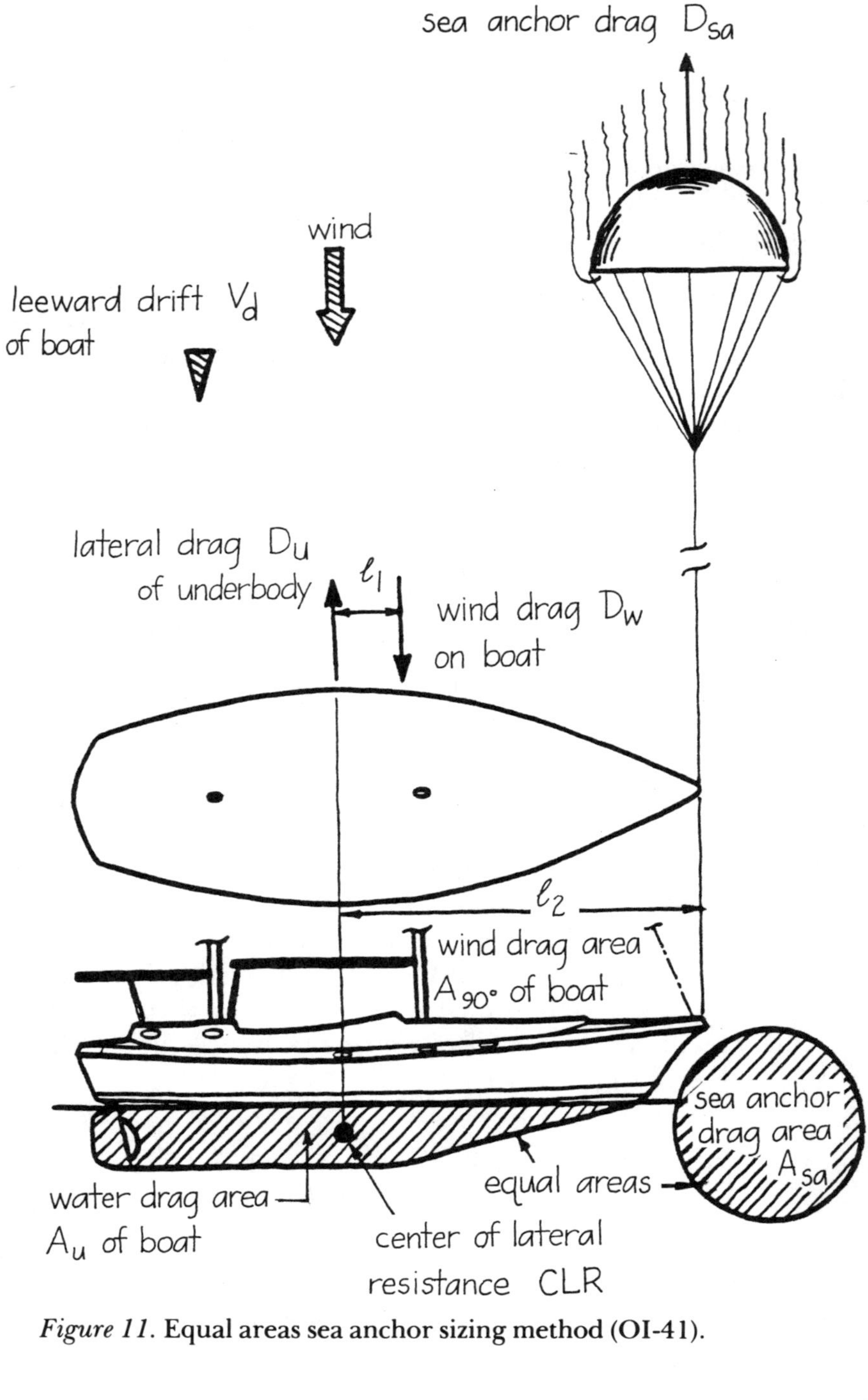

Figure 11. **Equal areas sea anchor sizing method (OI-41).**

$$A_u = 103 \text{ square feet}$$

Therefore:

$$d_{sa} = \sqrt{\frac{4 \cdot 103}{\pi}} = 11 \text{ feet}$$

This is a substantial diameter, necessitated by the long keel which actively tends to resist turning as the sea anchor brings the bow of the boat into the wind. Once the bow of the boat is head to wind, a sea anchor diameter of only 9½ feet (based on the wind drag method) would be required to hold it there provided some mizzen sail is set to help weathercock the boat into the wind.

Average Draft Method

Shewmon, Inc., recommends a special method for determining the diameter of their sea anchor for use on a modern fin keel sailboat. Since fin keel boats can pivot so easily about their narrow deep keel, the sea anchor needed to turn them into the wind can be smaller than that required for a long keel boat. In equation form the Shewmon recommendation is:

$$d_{sa} = 2\left[\frac{\Delta A_k}{LWL/2} + \text{hull draft}\right] \qquad [\textit{Equation 8}]$$

where d_{sa} = diameter of Shewmon sea anchor (feet)
ΔA_k = broadside area of fin keel (square feet)
LWL = waterline length (feet)

Example: The J/41 light displacement sailboat shown in Figure 12 has the following applicable dimensions:

ΔA_k = 31 square feet
LWL = 36 feet
hull draft = 1.8 feet

Therefore:

$$d_{sa} = 2\left[\frac{31}{36/2} + 1.8\right] = 7 \text{ feet}$$

If the sea anchor diameter had been determined by the equal areas method, it would have been 10½ feet in diameter—unnecessarily large. If the sea anchor size had been determined by the wind drag method, it would be only 5 feet in diameter because of the streamlined nature of this racing boat.

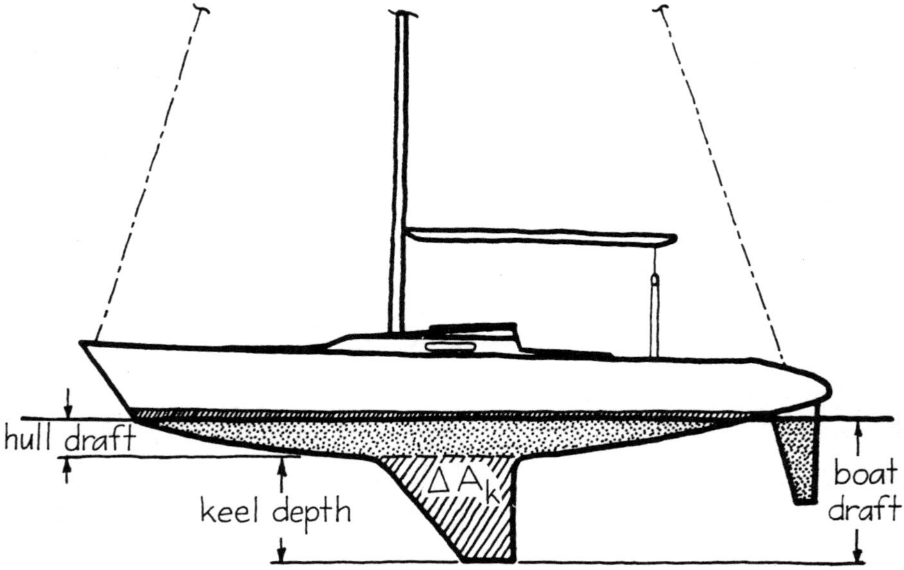

Figure 12. Profile of a light displacement ocean racing sailboat (J/41).

Shewmon, Inc., in published material dated April 1987, suggests using the following sizes of their sea anchors based on the average draft of the boat as determined by the preceding method:

Boat draft (feet)	*Sea anchor diameter (feet)*	*Tether diameter (inches)*
2	4	9/16
3	6	¾
4	8	⅞
6	12	1⅛
7½	15	1¼
9	18	1⅜

The manufacturer should be consulted before ordering any sea anchor, so that he can match a sea anchor size to your specific boat's configuration.

Van Dorn's Method

Another method for estimating sea anchor size was developed by Van Dorn in his book *Oceanography and Seamanship.* He offers an approximate solution to the sizing problem based on the waterline length of the vessel (assumed to be a monohull). His "optimum" sea anchor size is related to the wind speed at which you feel the need to sea anchor. At a wind speed of 30 knots Van Dorn suggests a sea anchor diameter equal to 16 percent of the waterline of the boat. This increases to 25 percent of the waterline for winds of 60 knots. Table 4 is a numerical representation of his arguments for drift speeds of less than 2 knots.

You can make your selection from this table based on the maximum winds that you expect in your area of operations. Obviously, if you design for the lower wind speeds, your sea anchor will be smaller and less expensive but not as good insurance as a large sea anchor. Your navigator should note that the drift speed V_d is relative to the water, which itself may be moving and must be taken into account in dead reckoning. It is recommended that any boat that operates in blue water have a sea anchor capable of handling 60 knots of wind.

Sea anchor size is determined in the Van Dorn method by the formula:

$$d_{sa} = (f_{LWL})\,LWL \qquad [\textit{Equation 9}]$$

where d_{sa} = uninflated diameter of sea anchor (feet)
f_{LWL} = ratio of sea anchor diameter to waterline length (Table 4)
LWL = waterline length (feet)

Example: A sea anchor is desired for the Out Island 41 medium displacement sailboat to handle winds of 60 knots. What should its size be?

$LWL = 34$ feet From Table 4 $f_{LWL} = 0.25$

Therefore:

$$d_{sa} = 0.25 \times 34 = 8\tfrac{1}{2} \text{ feet}$$

f_{LWL} (ratio of uninflated sea anchor diameter to waterline length)	Wind speed V_w (knots) 30	45	60	
0.16	1.3 kts	2.1 kts	2.7 kts	Not recommended due to too high a drift speed
0.21	1.2	1.6	2.2	
0.25	0.9	1.3	1.8	Van Dorn's optimum size recommendations

Table 4. Relative drift speed V_d based on Van Dorn's optimum sea anchor size (exclusive of current).

This is smaller than either the equal areas method (11 feet) or the wind loads method, which yields a 9½-foot diameter.

If we look at the J/41 light displacement sailboat, we see that it has the following applicable dimension:

LWL = 36 feet

Therefore:

$$d_{sa} = .25 \times 36 = 9 \text{ feet}$$

This is considerably larger than that calculated by either the average draft method (7 feet) or the wind drag method (5 feet). It would seem that Van Dorn's method is less applicable to the new breed of light displacement sailing craft.

After reviewing these several methods of estimating sea anchor size, you are still left with the dilemma of deciding which method will give the best estimate for your boat. There is no single method best suited to all boats, but if you look at your boat's predominant design feature, you can use the method that best takes it into account. For example:

Design feature	*Matching method*
Long-keeled boats: monohull sailboats, displacement and deep-V powerboats	equal areas (conservative) or wind drag
Flat bottom planing powerboats	wind drag
Fin keel sailboats	average draft

MAKING YOUR OWN SEA ANCHOR

Making your own sea anchor is not impossible, but the results may be less than desired unless an extensive testing program is carried out before the sea anchor must actually be deployed in heavy weather. Three critical factors must be considered: size, stability, and strength. You have already been taken through the steps to determine the load which the sea anchor must handle. For the example using the Out Island 41 boat, the steady state wind load was 2,650 pounds. We can assume that surge loads will double that figure to about 5,300 pounds (see my book on anchoring [Hinz 1986] for a rationale of added surge loadings). Now we put a 100 percent safety factor on the sea anchor assembly and end up with a total design loading of 11,000 pounds. (The tether must also handle this same load.) These loads are distributed through the shrouds connecting tether and canopy. It is easy to see that even the 9-foot-diameter sea anchor can develop some pretty tremendous loads.

Canopy materials should be heavy synthetic cloth; Dacron or nylon are preferred candidates because of their all-around durability. Material weight should be between 3 and 5 ounces and the sewing should conform to the best sailmaking standards. Puckers and gathers should be meticulously avoided. Opening shocks of a poorly deployed sea anchor and surge loadings could split poor seams, weakening the canopy assembly.

When you design the rope or webbing connections for the sea anchor, make allowances for the following strength reductions:

Bowline	60% of rope strength		
Anchor bend over a ⅝-inch-diameter ring	75%	"	"
Eye splice with thimble	90-95%	"	"

Webbing, too, suffers a marked reduction in strength due to being folded and having stitching penetrating its body. Parachute makers' tests have shown that folded and sewn webbing has an assembled strength of only 70 percent of the virgin webbing. Allow for this in your design.

Stability of the canopy under load depends mostly on the ratio of shroud length to canopy diameter. As an absolute minimum the ratio should never be less than one; even then, such a short shroud cone will cause the inflated diameter of the sea anchor to be held to about 70 percent of the uninflated diameter, a significant loss of drag area. A shroud length of over 2 canopy diameters appears unnecessary from a canopy opening standpoint since longer shrouds do not allow the canopy to open any farther. Instead, they tend to cause an instability in the sea anchor. Long shrouds promote canopy rotation, resulting in a shortening of the shrouds and entanglement of the trip line if one is used. There is also a greater tendency for the sea anchor to oscillate in the water when longer shrouds are employed. A shroud length ratio of 1.25 to 1.5 times the canopy opening diameter is a good place to start with your design.

There is really no proven way to test your homemade sea anchor's performance short of heavy weather use. The Shewmon company's sophisticated testing system using 1,500-horsepower harbor tugs is well beyond the resources of the person who feels it economically expeditious to make his own sea anchor. The only alternative is for the skipper to take his boat out in the heaviest weather deemed safe for a test and actually deploy the sea anchor without an emergency being the driving force behind the deployment. The sea anchor can then be observed for adequate drag (at that wind speed), stability, strength (at that wind speed), and general behavior in deployment and retrieval. Such an experience

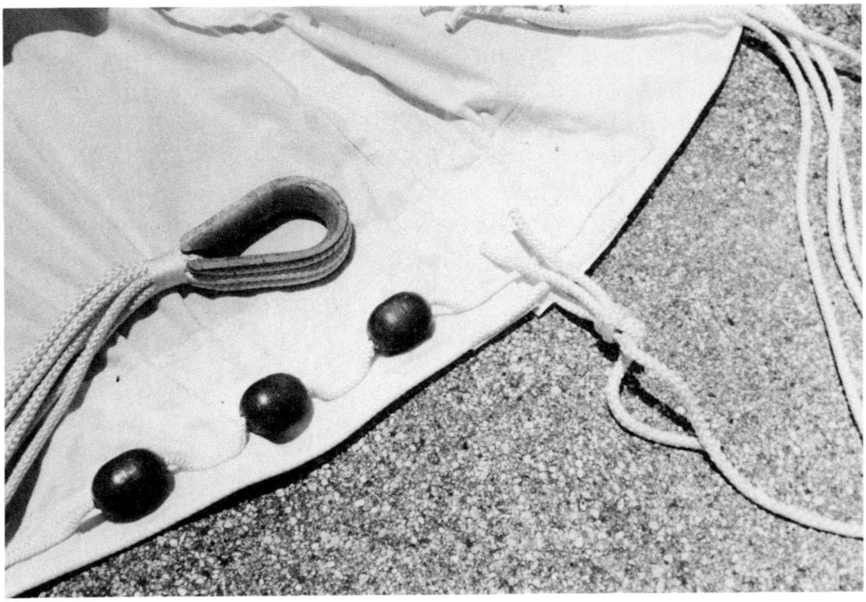

Small weights are sewn to the hem of the Shewmon sea anchor to prevent rotation. Courtesy Shewmon, Inc.

will pay off later when the sea anchor must be deployed under more extreme and adverse conditions.

JURY-RIGGED SAIL SEA ANCHORS

In an emergency sailboats can make use of a sail as an improvised sea anchor. Low aspect ratio sails work the best, but any will do in a pinch. Simply attach shroud lines to tack, clew and head, making them of such length that they simulate equal length shrouds of an equivalent circular sea anchor as shown in Figure 13. Gather the shrouds together and connect them to a tether and your sea anchor is ready to go. Do not be concerned about having too large a sea anchor due to the size of the sail at hand. A large sail will put no more strain on the boat than a smaller one. Remember the ground anchor analogy.

Shewmon points out that the sail sea anchor is not as efficient as the built-for-the-purpose circular sea anchor. If you have used a gaff-rigged sail, then the efficiency is about 70 percent. For a triangular sail, it is about 50 percent. Also, a sea anchor made out of a sail will tend to be unstable and will yaw back and forth as the water spills alternately from one edge or

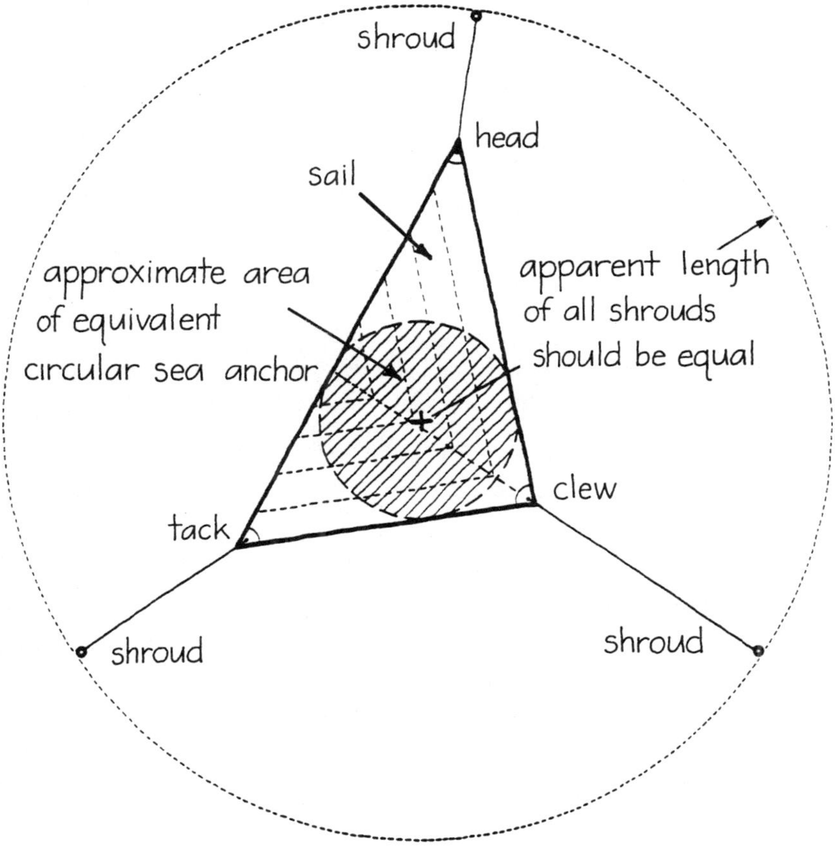

Figure 13. Using a sail as a sea anchor.

the other. But it will furnish drag and that is what you are after in an emergency.

There is no simple way to trip a sail sea anchor, so you will have to move the boat up to it when the emergency is over, either using the motor or by winching in on the tether. The sail should be thoroughly washed afterwards with fresh water (the next rain?) before being dried and stowed away. Check it for damage at the clew, tack, and head fittings and repair as necessary. Now you have time on your side.

FOUR

Sea Anchor Use

The components that make up a sea anchor system are similar to those of a ground anchor system, but with some different names:

Ground anchor system	*Sea anchor system*
deck gear	deck gear
rode	tether
ground anchor	sea anchor
trip line (in some cases)	trip line (in some cases)

You use the same deck gear for a sea anchor as you use for a ground anchor. The reason is obvious—it is there and it is sized for the job of "anchoring." The ground anchor rode can be used as part of the sea anchor tether provided that it is an elastic rope rode of sufficient length. The design of a sea anchor tether is critical to its success. A sea anchor tether of improper length, inadequate strength, wrong construction, or incompatible components can doom the vessel it was intended to aid.

RIGGING THE SEA ANCHOR

The components of a fully rigged sea anchor are illustrated in Figure 14 (a). The monohull tether is brought aboard the boat in the same manner as a ground anchor rode—through a well-faired bow roller with plenty of chafing gear. Only one swivel is possible in the monohull tether and that is placed at the apex of the shroud lines of the sea anchor.

Multihull sea anchor tethers must be brought symmetrically to the hulls through the use of a bridle, as illustrated in Figure 14 (b). The key to the use of a bridle is to have the included angle less than 30° in order to minimize the lateral loading put on the hulls. Bridle lines should be run through snatch blocks and then back to cleats so that the symmetry of the bridle as well as the overall length of the combined tether-bridle can be adjusted as the wave length of the seas changes. It is common practice to make the size of the bridle lines equal to that of the tether although one size smaller is adequately strong.

The bridle on a multihull provides a benefit not found on the single tether of the monohull and that is it holds the multihull into the wind better, minimizing the amount of yawing that takes place. The bridle swivel not only brings the lines together but it helps prevent rotation of the nylon tether.

Sea anchor swivels, like those of a ground anchor, should be quality forged steel or bronze of an adequate size. The swivel should be inspected before sea anchor deployment to be certain that it is free to do its job with a minimum of friction.

A trip line can be attached to the crown of the sea anchor to help in the final retrieval of the sea anchor. The trip line and the trip line extension can be made of ⅜-inch-diameter nylon or Dacron in braided construction. A floating trip line would be preferred but polypropylene rope tends to kink and tangle too easily to be used for this purpose. One of the newer synthetic ropes that has good strength and elasticity properties and floats as well may be the best answer although its cost is a bit high.

The sea anchor float can be any docking fender, but one of the spherical, fluorescent floats is the best for visibility. When the sea anchor is deployed, it is this float that tells you how far upwind your sea anchor is, allowing you to position it one wave length (or multiple thereof) upwind of your boat. A small swivel is also needed at the juncture of trip line, trip line extension, and float eye. The trip line extension length should be 1½ to 2 times the height of the largest wave to keep the sea anchor below the main orbital motion of the wave particles.

The fabrication of the tether assembly should follow the best practices of putting together a ground anchor rode: tight eye splices around proper thimbles which are seized in place, safety-wired shackles, freely moving swivels, chafe protection where needed, and great care to prevent tangling of lines in stowage.

During use, chafe prevention is your number one worry. Generous size cleats, large radii fairleads, and leather or corded hose chafing gear must be used. All the while the sea anchor is deployed, its attachment to the boat must be inspected frequently and, if chafe is developing, a fresh nip must be taken on the line.

The most severe chafe on the tether will probably take place where the tether passes over the bow roller or chock. This can be alleviated through the use of a length of chain attached to the braided nylon tether.

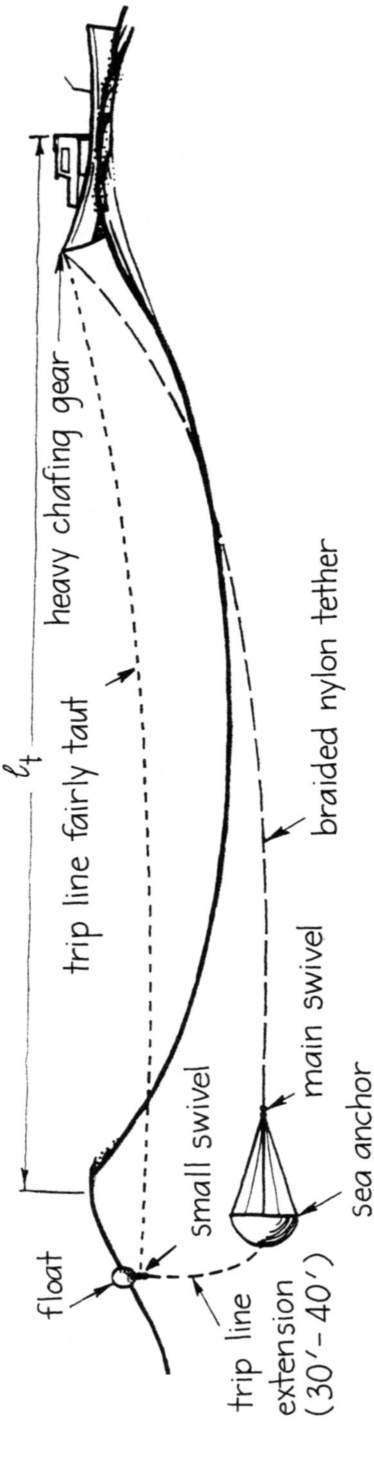

(a) Monohull attachment

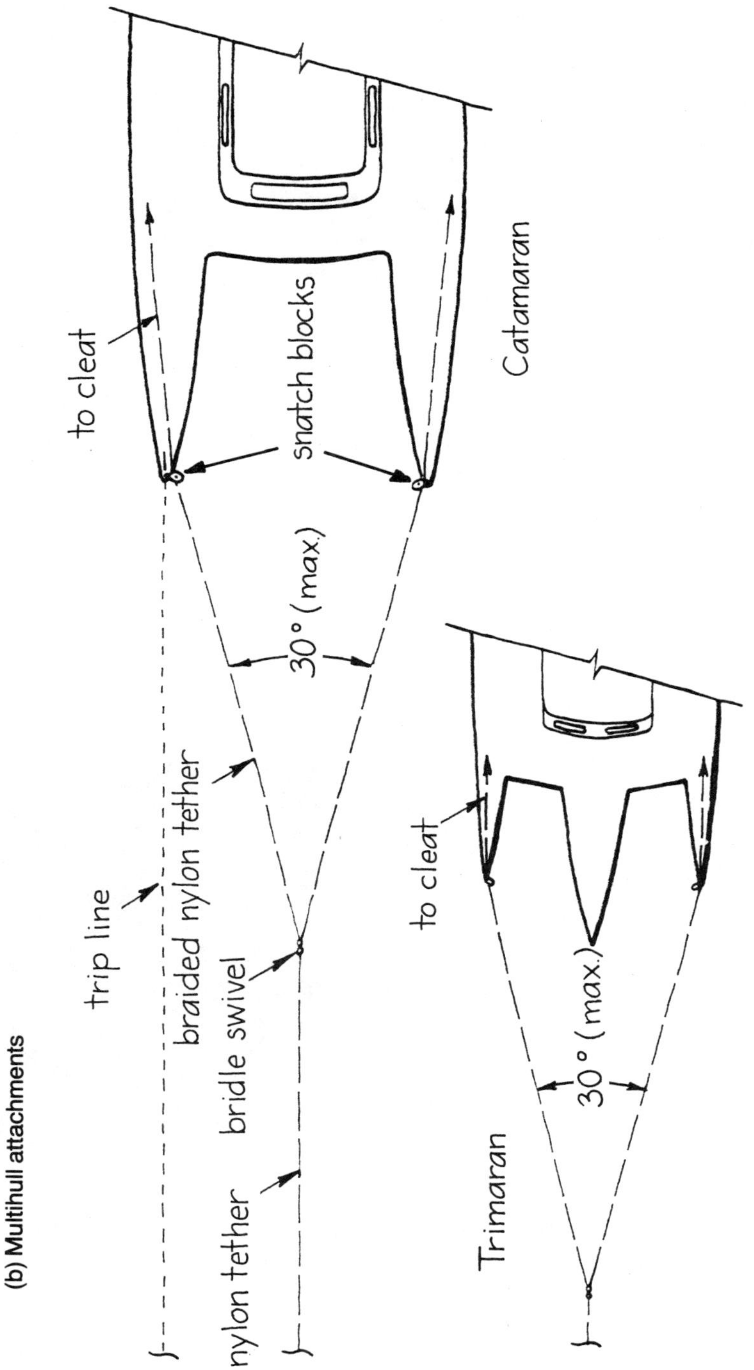

Figure 14. Rigging the sea anchor.

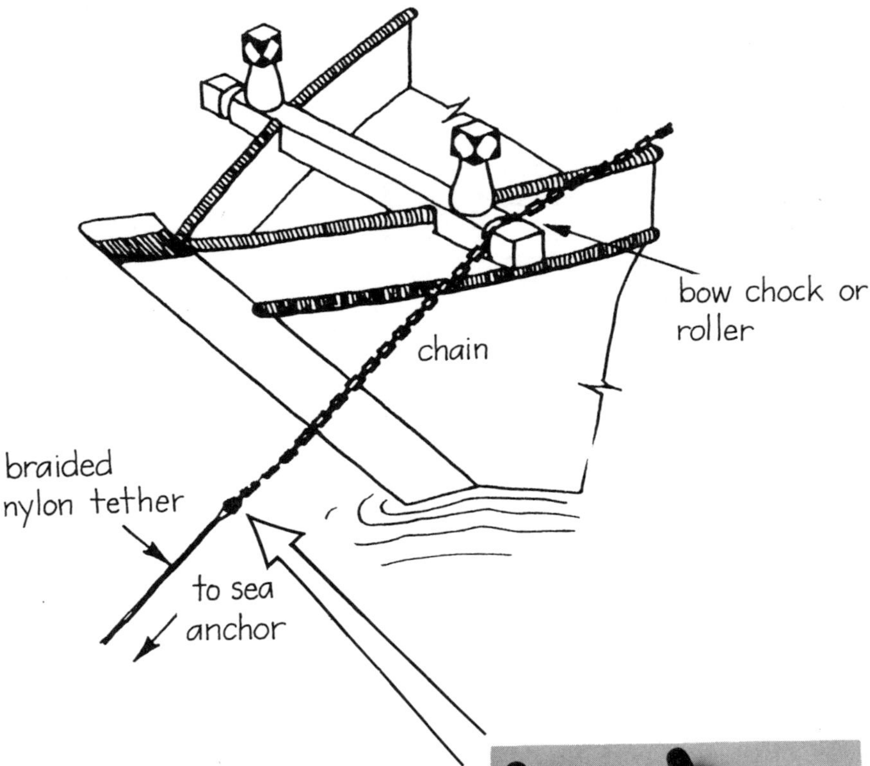

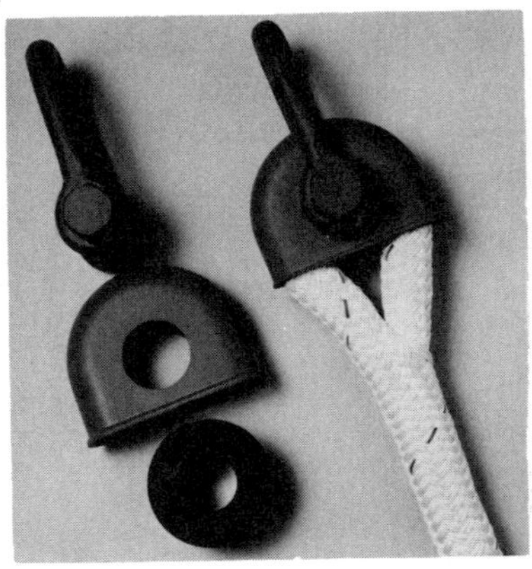

The Samson Nylite connector is a 3-part assembly: a spool, a shield, and a special shackle. It can be used with either braided or twisted synthetic rope. This is a major advance in the technology of linking rope to chain. Courtesy Samson Ocean Systems, Inc.

Figure 15. Eliminating chafe at the bow roller. The illustration shows a recommended way to connect a chain to a rope tether to prevent chafing of the rope. The boat in this case is a fishing sampan fitted with catheads for general ground tackle use. Other boats may use bow rollers or chocks to handle anchor rodes.

The attachment should be made with a non-chafing connector such as the Samson Nylite connector (Figure 15) or other thimbles for storm anchor rodes shown in my book on anchoring and mooring (Hinz 1986). This solution requires that the rope rode be made a fixed length and that any length adjustment be made by letting out additional chain. A good starting length for the braided nylon rode would be about 200 feet with the option of adding up to 100 feet of chain. There is a limit to the amount of chain that can be added since the weight of the chain will tend to sink the sea anchor too far. This scheme is illustrated in Figure 15.

TETHER LENGTH

Many (if not most) references relate the tether length (ℓ_t) to some multiple of the boat's length, say 10 boat lengths, or to some arbitrary length such as 200 feet. Neither is correct although there may be an occasional set of waves where each length becomes appropriate by chance. Rather than resorting to chance, however, one can use a clear rationale to determine tether length. This is based on the wave length of the sea for the particular wave condition at hand.

Consider the relation between adjacent wave crests, the sea anchor, and the boat as shown in Figure 16. When both sea anchor and boat are riding on crests (or in troughs) as in Figure 16 (a), they experience the same orbital motions of the surface water and, hence, move in harmony with each other. There is no added acceleration loading due to differential speeds. In this case the tether length ℓ_t is nominally equal to the wave length L.

If the tether length is less than the wave length (or greater, also), then the boat may ride a crest while the sea anchor is in a trough as shown in Figure 16 (b). Relative to the boat, then, the sea anchor will be accelerated forward while the boat will be accelerated aft. The result in all probability will be a torn sea anchor body, a failed tether, or the tether attachment on the boat being torn asunder. Whatever happens, the boat will then be left to broach, roll, or pitchpole, none of which is an attractive alternative to providing a suitable tether length in the first place.

If the boat is riding a trough while the sea anchor is riding a crest as illustrated in Figure 16 (c), then the tether will experience a complete slackening with the possibility of the sea anchor being caught up in the wave overfall. This could lead to the shroud lines becoming entangled and the sea anchor not reopening. The consequences of that happening need not be stated.

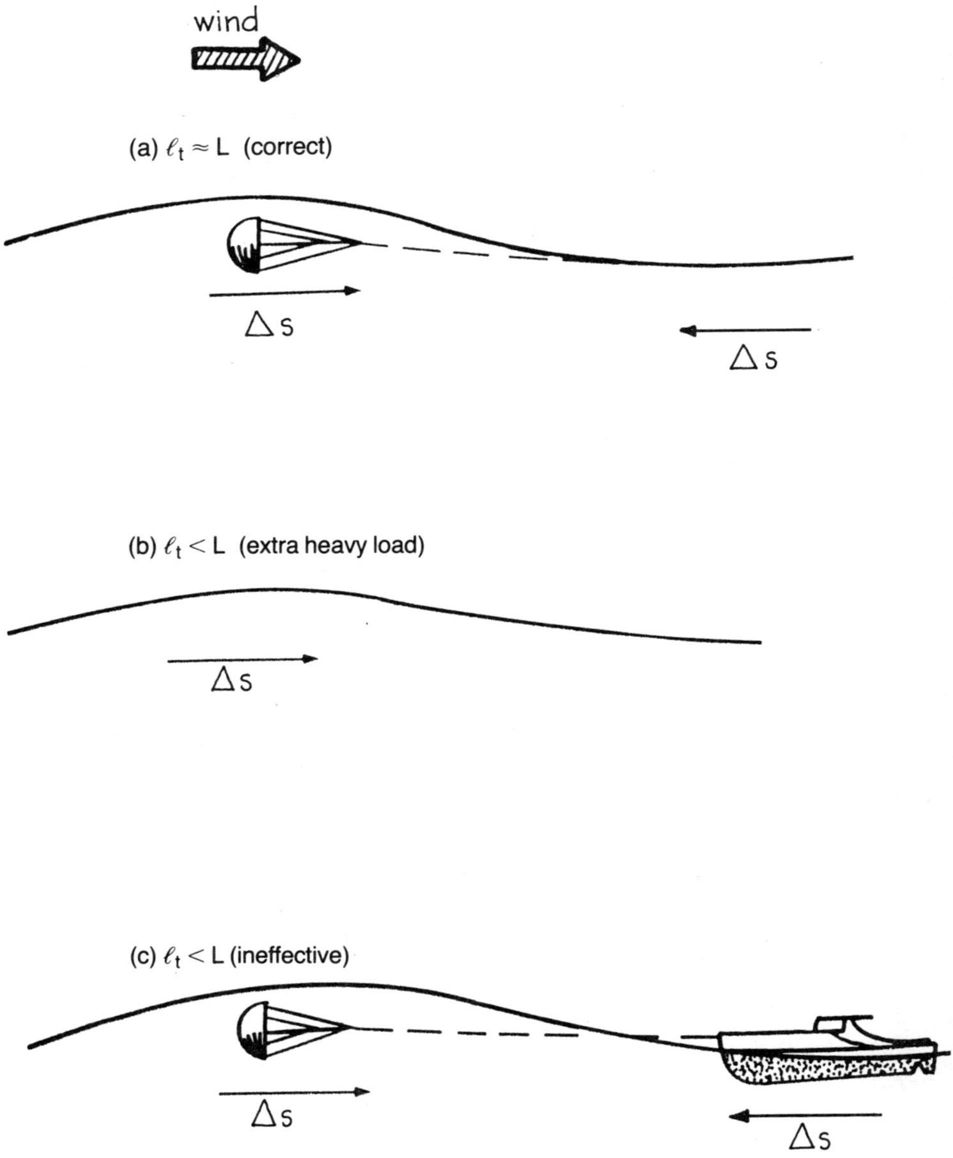

Figure 16. Matching tether length to wave length.

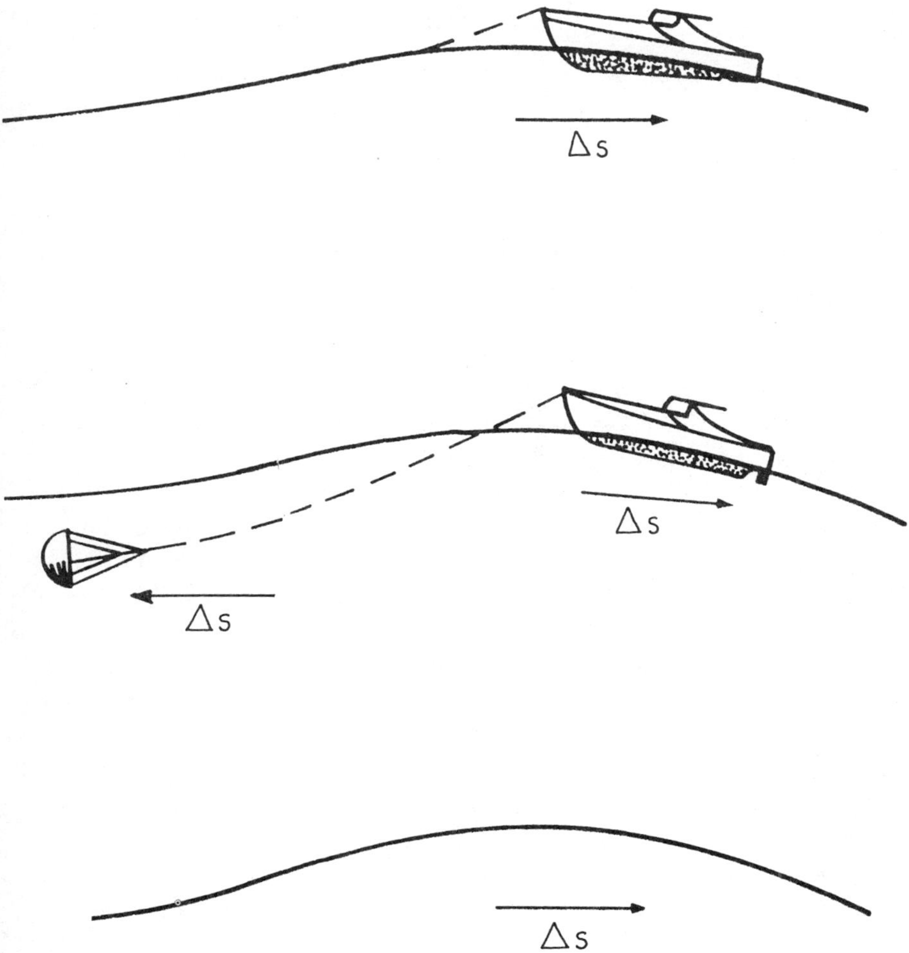

Δs
Δs
Δs
Δs

It is essential that the length of the sea anchor tether be matched to the wave length of the sea (or a multiple thereof) for a particular wind condition. And you should be able to adjust it to match significant changes in wave length. What, then should the actual length of the tether be?

Figure 17 has been developed to give the practical tether length for various winds as measured by Beaufort number and for various durations of blowing of those winds. An unlimited fetch has been assumed. According to Bowditch, the actual wave length of a sea is about two-thirds that calculated by Equation 1 and this has been plotted in the figure as the tether length. Note: The tether length varies with both the duration of the blow and the intensity of the blow, hence the need for an adjustable tether.

Determining the proper length of a sea anchor tether in seas raised by a steady state wind is fairly easy using the foregoing method, but when it comes to seas raised by shifting winds, the solution is not so clear. Confused and disorderly seas are the norm for gales that approach from a direction different than the prevailing wind. In this case it would be hard to identify the critical wave length by Figure 17. Under these conditions, then, you must resort to an "eyeball" length determination. Assuming that you have a highly visible fluorescent docking fender as a float, adjust the tether length to match the wave length of the dominant wave form.

A point should also be made regarding the length of the tether to be used in a shallow body of water, such as one of the Great Lakes. The same principles should be followed; that is, the tether length should be an integral wave length long. But, because one wave length in shallow water may be quite short, the benefit of elasticity of a long tether is lost, so consideration should be given to making the tether multiple wave lengths long, thereby preserving the elasticity while still matching wave lengths to minimize jerking.

When the tether of a parachute-type sea anchor goes slack, the sea anchor simply relaxes and floats in the water since there is no speed differential with the water. It will have no tendency to tumble or turn itself inside out unless it operates too shallowly and gets caught in the overfall.

TETHER STRENGTH

The desired strength of the tether is not amenable to simple mathematical solution. There are simply too many unknowns in the equation. As seen in Figure 18, we have at least five drag components to

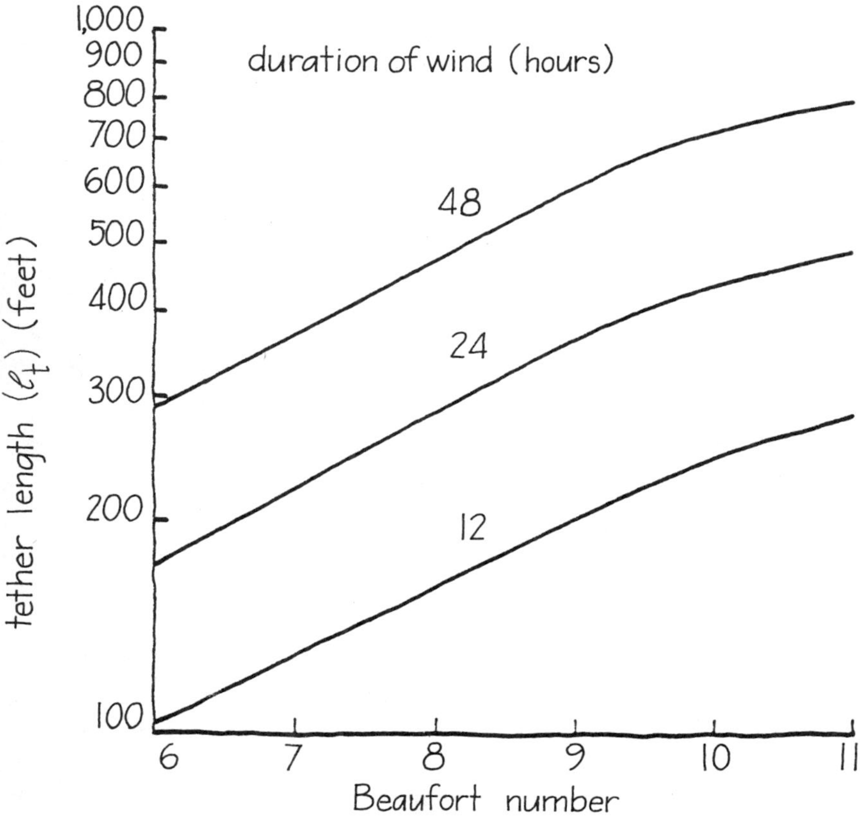

Figure 17. Length of sea anchor tether.

consider, of which only one, the wind drag, is anywhere near definable in a rational manner. The drag due to crest overfall is highly dependent on the intensity of the wave break. The drag due to the boat's tendency to backslide down the face of the wave depends on what pitch angle the boat assumes at any instant in time. The inertial drag of the boat depends on the relative orbital speeds of the sea anchor and the boat, as well as the elasticity in the tether and the ability of the boat to accelerate forward with the sea anchor.

To untie this Gordian knot, we must resort to practical experience, which also helps us with the logistics of our sea anchor gear. Most instances of sea anchor deployment have seen the use of the ground anchor rope rode as the tether. This may have been an expedient in those cases,

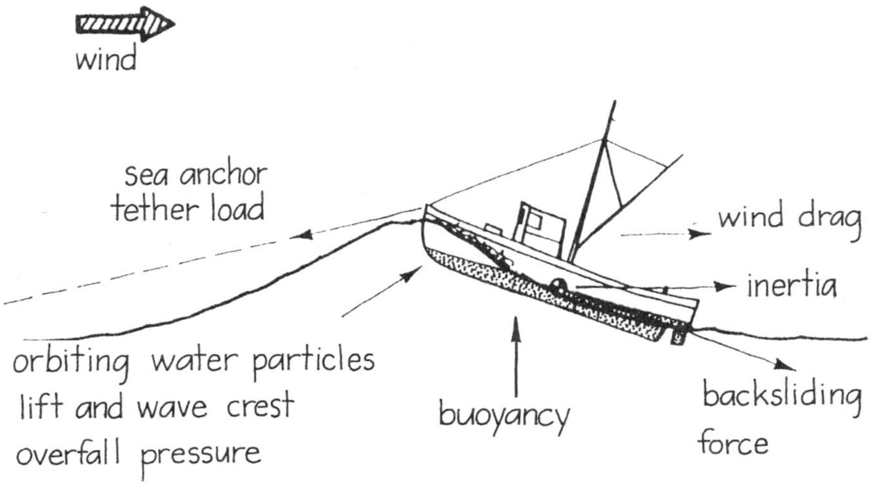

Figure 18. Forces on a boat riding to a sea anchor.

but it has worked successfully and it is available. Maybe the tether need not be of such size, but to make it smaller runs the risk of promoting failure, within the limited knowledge that we have. Until some boat fits a load-measuring system in a tether and goes out in storm seas, we had better stick with what practical experience has told us.

You can calculate the size of the rope anchor rode for your boat by one of the two methods described in my book on anchoring and mooring (Hinz 1986). The length of a tether may need to be considerably longer than a ground anchor rode (meaning that your boat must carry extra line on board of the proper size and there must be a means of connecting lengths together in a reliable manner). Such a scheme is suitable for the coastal sailor who will try to avoid stormy seas whenever possible. For the blue water passagemaker, however, it would be prudent to carry on board a made-for-the-purpose sea anchor tether.

The tether, just like a ground anchor rode, must have elasticity to absorb surge loads. For this reason nylon rope is the best choice. But the type of construction is not so clear. One of the characteristics of a sea anchor is to rotate if it has a chance. When 3-strand rope construction is used, there is every opportunity for it to rotate even though swivels are used in the tether. A rotating sea anchor can entangle the trip line and render the canopy ineffective. To avoid this, braided nylon rope may be preferred even though there is some loss of elasticity. This provides the nonrotating feature as well as some elasticity to cushion wave impacts on the boat. The strength comparison of these two forms of rope construction is shown in Table 5. Braided nylon is 25-35 percent stronger than twisted construction with a higher allowable working load. Its elasticity, however, is only 60 percent that of twisted construction, as seen in Figure 19. For this reason, if you substitute braid for twist, match them on the basis of breaking strength and not diameter. That way you retain maximum elasticity in a tether of smaller diameter.

Writers have commented on tethering the boat too firmly. One says, "A possible danger in using a sea anchor is that the boat might be tethered or held too firmly against the seas so that she cannot yield," and "Even if the modern boat can be made to lie reasonably well to a sea anchor forward and a riding sail aft, there are certain disadvantages in this method, a major one being that the boat will at times be thrown backwards, and this could very possibly damage the rudder."

The question of holding the boat too firmly is answered by the elasticity of the tether which, as in ground anchoring, stretches to relieve the surging load. That is much better than having a too-small sea anchor which allows the boat to drift aft at such speed (or acceleration) that the rudder is put in jeopardy.

As for the boat being thrown backwards, that is not likely to happen if the proper length of tether is used such that both boat and sea anchor ride in fields of orbiting water particles traveling in the same direction. This is

Diameter of rope		3-strand twisted nylon		Braided nylon	
inches	millimeters*	Breaking strength (pounds)	Recommended working load† (pounds)	Breaking strength (pounds)	Recommended working load‡ (pounds)
3/8	9	3,700	410	4,800	960
7/16	10	5,000	550	6,500	1,300
1/2	12	6,400	700	8,300	1,660
9/16	14	8,000	880	11,200	2,240
5/8	16	10,240	1,140	14,500	2,900
3/4	18	14,200	1,560	18,000	3,600
7/8	22	20,000	2,200	26,500	5,300
1	24	25,000	2,750	31,300	6,260

Table 5. Strength of ropes (dry).

*Closest metric rope size.
†Working load is 11% of average breaking strength.
‡Working load is 20% of average breaking strength.

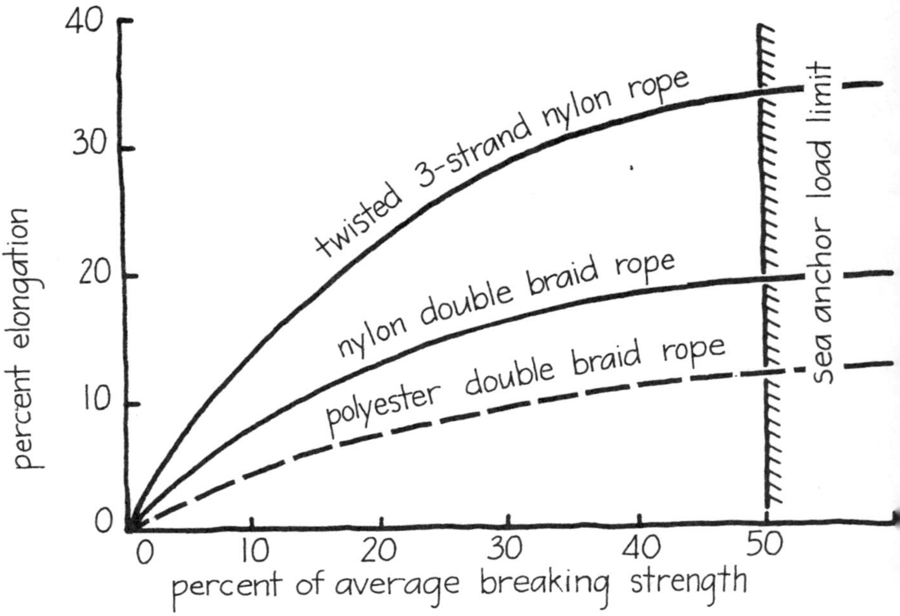

Figure 19. Elongation characteristics of synthetic rope.

critical and the bulk supply of tether carried on board should be long enough to permit lengthening it should the wave length further increase with an increase in the wind.

DEPLOYMENT

You are probably wondering by now how to deploy the sea anchor. There is only one answer to that—with great care. If you are deploying in a storm, you will have strong winds and seas to contend with and getting any part of your sea anchor hung up on the boat's rigging or fittings could be disastrous.

Deployment of the parachute sea anchor takes place off the windward side of the drifting boat or, if downwind control can be maintained, the sea anchor can be deployed over the stern, but the tether is still attached to the bow. The procedures recommended by Para-Anchors International (Shane 1983) can be generalized as follows:

1. Head the boat into the wind and furl the headsails in a seamanlike manner while wearing a safety harness.

2. Secure the rudder amidships and retract centerboard(s), daggerboard(s), swing keel, etc. Furl mainsail and/or mizzen sail(s) in a seamanlike manner.
3. Deposit the parachute in the sea from the windward side near the bow.
4. Allow the boat to drift away from the parachute, snubbing up on the tether occasionally to help the parachute to open and to rotate the bow of the boat into the wind.
5. Continue to drift back from the parachute, paying out the required length of tether. Keep some tension on the tether to prevent the parachute from sinking too deep.
6. When the proper length of tether is deployed (boat and parachute are riding on the same parts of the waves), make fast the line and install the necessary chafing gear.
7. Secure the rest of the boat and take a coffee break.

If your sea anchor is fitted with a float and a trip line, they should be laid out on deck in the order in which they will be deployed: float, canopy, trip line and tether in parallel, and then the bridle if one is used. Ascertain that all shackles are safety-wired and that the bitter ends of the tether and the trip line are attached to the proper hull fittings. The entire assembly should be able to lift off the deck without catching on any of the boat's gear or tripping the crew.

The float is set over the side first, followed by the canopy and initial lengths of the tether and trip lines. When the canopy has filled, let the movement of the boat away from the canopy pull the remaining tether and the trip line off the deck. This may happen fast, so be certain there is nothing to snag the lines in their haste to go over the side. When the tether reaches its bitter end, the boat will turn quickly into the wind and assume the bow-on riding position. Should any deployment foul-up occur, secure the trip line short until it can be cleared. Then give the canopy its freedom again and finish the deployment.

A story ran in *Yachting* magazine (May 1987) that indicates some of the trouble that you may experience in deploying a sea anchor. The 76-foot Palmer-Johnson motorsailer *Scheherezade* had departed New England in December 1986 on a passage to Bermuda when it ran into some of the Atlantic's worst weather. The story reads:

> With 30 to 40 foot seas, thunder, lightning, heavy rain and hurricane force winds blowing, it was almost impossible to distinguish the ocean from the heavens. At 0200 the winds suddenly dropped from 70 to 30 knots, then to zero—the eye of the storm, apparently. Our hopes brightened until a couple of hours later, when the winds picked up again to 50, 60, then 75 knots. Suddenly the stainless-steel staysail halyard broke, and the staysail blew out. With no engine, no sail, no speed, the boat began drifting sideways. The large roller-furling jibs were unuseable in this kind of weather, and there was no spare staysail halyard. We put out a parachute sea anchor, but its 300 foot, one-inch nylon line got fouled in, or around, the propeller/rudder and obviously got severed by the knives meant for cutting fishing lines loose.

(The *Scheherezade* eventually was returned to a New England port under Coast Guard tow.)

Riding to a sea anchor, the boat will have two mechanical problems to overcome. One is with the rudder and the other is with the propeller. While tethered to the sea anchor, the boat will slowly move astern and it will also occasionally fall astern as the tether alternately snubs up and loosens as waves pass under it. To protect the rudder, it should be firmly lashed amidships before the anchor is deployed. If the boat has a tiller, the tiller can be secured firmly in place. If the boat has a wheel, the problem becomes more difficult to solve, since locking the wheel may still leave a lot of spring in the cable system. In this case, the rudder should be provided with a hole at the trailing edge through which a securing line to each quarter can be attached. It is critical that the rudder be fixed in place to prevent it taking up a side load if the boat is thrown backwards.

Self-steering systems should be similarly locked in place or lifted from the water to avoid damage.

The propeller problem is one that is unique to folding propellers. As the boat surges fore and aft, the blades of the propeller will alternately open and close. It is unlikely that this motion would permanently damage the blades, but it will make objectionable noises in the boat (which are unneeded when tensions may be running high enough already).

Many monohulls will not ride steadily bow-on to the wind; instead, they will tend to sail off on one tack or another. This results in severe lateral loads being applied to the stem fitting where the sea anchor tether comes aboard. If the stem fitting is actually on a bowsprit, then the side loadings may exceed the sprit's strength. To counter this dual problem,

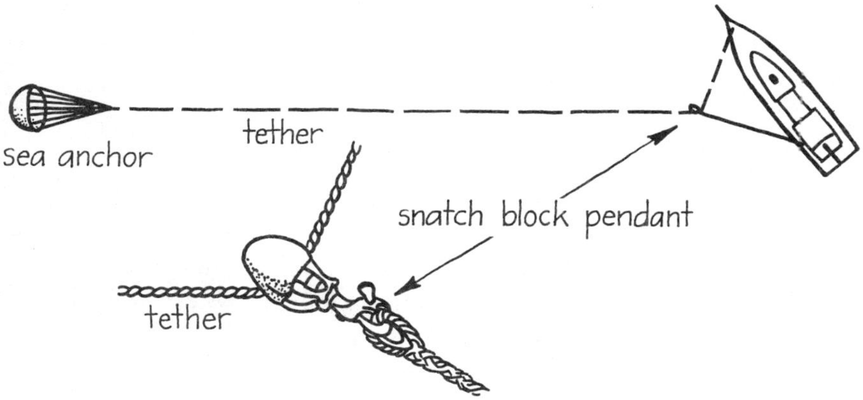

Figure 20. Pardey bridle.

your sea anchor tether can be rigged with a "Pardey bridle" which is simple and effective. The bridle shown in Figure 20 was developed by Lin and Larry Pardey while sailing their 24-feet LOD (length over deck) *Seraffyn* around the world (*Sail*, August 1982). It is essentially a barber haul applied through a snatch block to the sea anchor tether. The barber haul is belayed to a cockpit winch while the tether is belayed aft of the stem or sprit to a substantial cleat or Samson post. The angle at which the boat rides to the sea anchor and the wind can be adjusted to yield the best ride under the immediate conditions. As with ground tackle, chafe must be prevented at cleats and a fresh nip should be occasionally taken on the pendant to keep the line fresh.

Navigation

Navigation with the sea anchor deployed will be mostly a matter of tracking the current and estimating the downwind drift of the boat due to the slippage of the sea anchor. Both current and downwind slippage are about the same magnitude. In the open sea there is little to worry about, but around islands, coral, or other landmasses, a DR plot should be kept up to date to foreshadow any problems. Should your DR plot show that the boat is going to drift onto land, there are two choices, depending on what is known about the land characteristics. If it is a gradually sloping, good holding bottom, and the winds are not too strong, it may be possible to deploy your ground anchor to snub up while still offshore. If the shoreline is rocky, fast shoaling, and very deep, you had best retrieve the sea anchor while you have time and get the devil out of there. A boat not under control near land is certain to have a precarious future.

RETRIEVAL

Sea anchor retrieval would be very straightforward if it were not for the elements that forced you to deploy it in the first place—strong winds and high seas. If the winds and seas are down, retrieval can be easy, but if you are descending on a lee shore and need to retrieve your sea anchor during the storm, it can be difficult.

To retrieve a parachute sea anchor, haul in on the trip line, dumping the canopy. To do this you most likely will have to put the trip line on a winch or windlass because the loads will be quite high (and that is the reason for having a ⅜-inch trip line). Beware a springy nylon trip line, though, that is brought in at near-breaking strength of the line. If the line is taut and it should break while in the air, it will have a nasty snap back to the boat. If the line is immersed in the water when it breaks, in all probability the snap back will be totally damped by water friction. This would be a good reason to have a line with minimum elasticity as your trip line, e.g., a braided Dacron line reserved for the purpose.

A somewhat easier way to retrieve a sea anchor equipped with a trip line is to snub up on the trip line and then carefully ease out on the tether, allowing the trip line to upset the canopy as the tether goes slack. The collapsed sea anchor is then hauled in with the trip line.

A third method of retrieval can be used with any sea anchor whether or not equipped with a trip line. It is the obvious winching of the boat up to the sea anchor using the anchor windlass or some other winchlike device on board. When the main swivel has been brought on deck, grasp a shroud line and pull it aboard, dumping the canopy. The one problem with this technique can be found in the loads on the canopy due to the heavy weight of water in it. It may be necessary to hitch an auxiliary line to one shroud line by a rolling hitch and then winch the auxiliary line in until the canopy spills the water. Conversely, the auxiliary line can be simply attached to a strong cleat and the tether eased out until it goes slack as in the preceding method. Then the sea anchor assembly can be handed on board.

Needless to say, in any deployment or retrieval of a sea anchor, great care should be exercised. Crew members should be wearing safety harnesses and be well instructed in their duties. All lines should be free to run, and not entangled with each other or other parts of the boat. Should a tangle occur during deployment, retrieve the sea anchor, clear the impediment, and then redeploy more carefully. Do not become complacent about the possible loads on a sea anchor—they are high and they are dynamic!

CARE OF THE SEA ANCHOR

Inspection requirements for the sea anchor (or drogue) system fall into two categories—those associated with the tether and those associated with the canopy. In the case of the tether assembly (including bridle and trip line, if used), take the following actions as soon as possible after use. This can be done on board the vessel after the winds and seas have subsided:

Rinse out the synthetic lines with fresh water and dry them thoroughly. Do not allow sea anchors or drogues and their tethers to dry after being wetted with seawater. The resulting salt crystalline growth will cut the fibers. Sunlight should also be avoided.

Inspect lines for chafing or hockles (twisted construction). If any severe chafing exists or hockles have appeared in the twisted construction lines, replace the lines. Consign the old lines to making up dock lines or other less critical applications.

Examine twisted construction lines internally for broken or frayed fibers that appear to be fused together due to excessive heat generated by high loadings (as, for instance, around cleats or a Samson post).

Check thimbled eye splices in the lines for looseness or rusting of thimbles. Rust can be damaging to rope.

Clean each shackle or swivel in the tether and apply graphite grease to swivel. Be certain that swivels are free to swivel. Safety-wire all screw pins.

Inspection of the canopy assembly should cover the shroud lines, the canopy proper, and the trip line attachment, if any. Perform the following steps after each use. These, likewise, can be accomplished on the vessel's deck in calm seas and light winds. The latter is quite important in the case of a parachute sea anchor.

Rinse the canopy and shroud lines with fresh water and dry thoroughly.

Check each shroud from main swivel to canopy looking for abrasion damage. Check loop at swivel for secure fastening and examine stitching of line to canopy for looseness or breaks. If knots are used at either end of shroud, be certain that they are tight and that the free end is lightly stitched to the main rope to prevent loosening of the knot.

To check the canopy, start at the apex of one gore and inspect the gore material and radial seams from end to end looking for tears, snags, and loose stitching. Continue the same process until all gores have been inspected. Pay particular attention to the shroud line attachments.

Check the strength of the fabric by firmly pushing a finger into it; do not use the fingernail, especially on parachute canopies, as it could penetrate the fabric of a perfectly good canopy. Aged nylon, particularly that which has been exposed to the sun for long periods, should be carefully examined.

The sea anchor assembly can be packed after it has been rinsed, dried, inspected, and repaired (if necessary). The canopy is folded first, the shroud lines next, and, finally, the tether assembly. To fold the canopy, stretch it out along the deck, securing the apex to a cleat. Pick up each shroud line in turn and accordion pleat each gore until all gores have been folded and lie on top of each other. Do this when the winds are light, especially in the case of a parachute canopy. Then roll the canopy up, starting from the apex and continuing to the midpoint. Secure this bundle so that it can't unroll and then proceed to prepare the shroud lines.

Stretch the shroud lines out on deck and comb them with the fingers to remove kinks and overlaps. Be certain that the lines have not been inadvertently turned inside out; if they have, turn them back so that no line overlaps any other. Gather the lines together and snake them back and forth on the unrolled portion of the canopy starting at the canopy skirt attachment. Allow the main swivel to hang out one side and then roll the canopy and lines together starting at the skirt end and rolling towards the center to meet the previously rolled part. The completely rolled canopy can now be put into a bag for storage with the main swivel on top. The sea anchor bag should be stored in a dry location on the boat where it is protected from sunlight, spillage of any liquids, or possibility of mechanical damage.

The sea anchor tether after inspection and repair, if necessary, should be carefully coiled up like any valuable line on a boat, and hung up in a locker or sail bin where it is readily available and will not get tangled. When the tether is also used as the anchor rode, maintain it as described in my anchoring book (Hinz 1986) so that it remains a trustworthy line.

FIVE

Specialty Sea Anchors

LIFE RAFT SEA ANCHORS

All life rafts are equipped with an attached sea anchor. Coast Guard-approved life rafts for 4 to 25 persons require a second anchor to be carried, unattached, in the equipment bag. There are two jobs which the life raft sea anchor must do—one is to minimize drift and the other is to add stability to the raft.

Very few rafts have a capability for sailing under any degree of control and it is best to "anchor" them to the water and minimize the wind drift, accepting as inevitable whatever current drift may occur. This would also enhance rescue attempts since it would minimize the search area. A sea anchor capable of holding a raft to current drift speeds is small, indeed, as will be seen later.

The second function of the sea anchor is to provide stability to the life raft in high winds and seas. Several years ago the Cruising Club of Australia ran a series of tests on life rafts in the open sea and found that conventional rafts without a sea anchor were susceptible to overturning even in moderate seas. More recently, the British National Maritime Institute, with the cooperation of major British, Danish, and German life raft manufacturers and the Icelandic Directorate of Shipping, conducted extensive open sea tests (40 knots of wind and sea state 9 or better) of life raft stability. They found that the only rafts to capsize were those whose sea anchors failed and the crew was unable to attach another. These tests were so conclusive on the importance of the sea anchor to stability that design of the ballast bags became of lesser importance than previously thought (*Yachts & Yachting*, 12 March 1982).

Other conclusions reached were: the usual sea anchor made with a square fabric canopy has a strong tendency to tangle in its shrouds and become ineffective, but the conical type on a long tether sinks below the wave action and becomes extremely effective. The life raft with a conical sea anchor used in conjunction with water ballast bags was completely stable in winds to 78 knots. Without the sea anchor the rafts would capsize nine out of ten times in 55 knots of wind.

The final sea anchor design was a conical shape made out of a close mesh polypropylene netting. Later tests indicated this design to be completely successful in preventing raft capsize even when one raft lost all of its ballast bags.

The largest life rafts in common use today are 25-man rafts equipped with two sea anchors—one attached and the other in the survival pack. The sea anchor provided with a 25-person life raft is a vented spheroconical type with an opening of 24-inch diameter, a vent of 3-inch diameter, and a length of 18 inches. Six shroud lines of 45-inch length attach to the canopy at one end and a 40-foot-long, ¼-inch-diameter, braided nylon tether at the other. The tether in turn attaches to a D-ring on the end of the rectangular raft (see photos).

The adequacy of this sea anchor can be estimated by a simple calculation of the drag of the raft and the drift speed with the wind. Using the methods previously described in Chapter 3, a 60-knot wind load on the

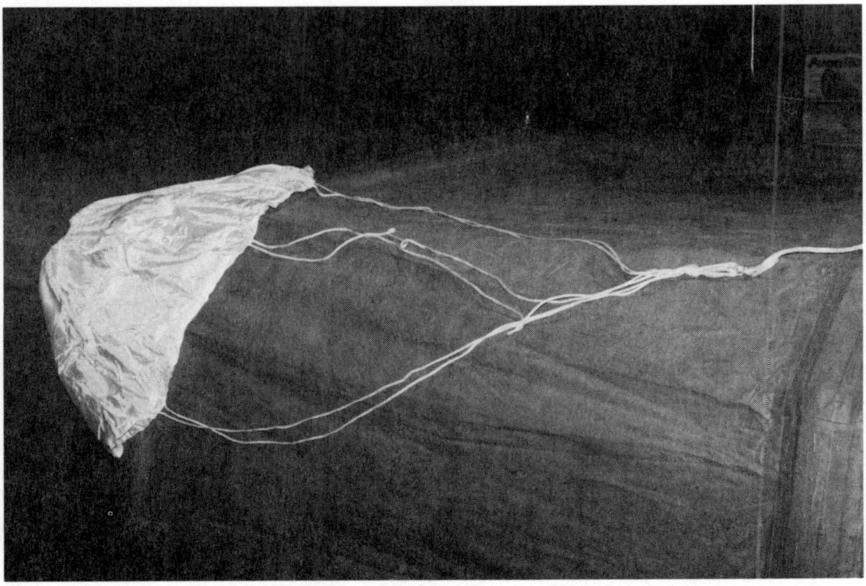

The 24-inch sea anchor that is standard equipment on a 25-person life raft. Photo by the author.

raft is approximately 150 pounds. The drift speed with the attached sea anchor would be about 1¼ knots in the 60-knot wind. The mass of water inside the sea anchor canopy that would have to be accelerated for a wave to overturn the raft is approximately 270 pounds. The ¼-inch tether has a breaking strength of 1,600 pounds so the sea anchor is capable of resisting a 5- to 6-G acceleration, thus providing substantial restraint to a wave overturning the life raft.

Scale model tests by Jordan (1984) substantiated the ability of a drogue to stabilize a life raft:

> Preliminary tests were conducted with models of the six man life raft (86 inch diameter full size with a 24 inch diameter disk-type drogue). In the 20 foot simulated breaking wave, the unballasted life raft with no drogue was violently tumbled. In a simulated 8 foot breaking wave, the raft would be capsized about half of the wave strikes. The provision of either a drogue or a hemispherical ballast bag prevented capsize in the 8 foot wave. In the 20 foot wave neither the drogue nor the ballast bag prevented capsize in most cases. However, with the hemispherical ballast bag, the raft would immediately right itself after capsize.

One life raft in the 1979 Fastnet Race carried eight men from the abandoned boat *Trophy*, but it was capsized several times by the wind getting under one edge of the raft and flipping it over. It did not have a working sea anchor. Two of the crew were lost in one of the capsizes. After the fifth capsize, the two flotation rings parted and the crew huddled in the lower ring which had the floor. They finally found the sea anchor, which had been so tightly wound up in its tether that it would not open by itself. The survivors cleared the tether and deployed the sea anchor over the side. It effectively slowed the raft's drift and steadied its motion, but soon afterwards the tether parted. The raft subsequently rolled over several times before the occupants were at last rescued (Rousmaniere 1980).

It becomes patently clear from this tragic episode that both wind and wave can overturn a life raft. It is further evident that care must be taken in deploying the sea anchor and that the tether must be durable enough not to chafe through. A properly deployed sea anchor can materially add to the stability of the life raft in severe seas.

One weak spot in life raft sea anchors is the tether's attachment to the flotation tubes. On one 25-person Coast Guard-certified life raft, the tether is knotted to a D-ring held on with a rubber patch in tension. This poses two problems: one is that the patch cement is not very strong in

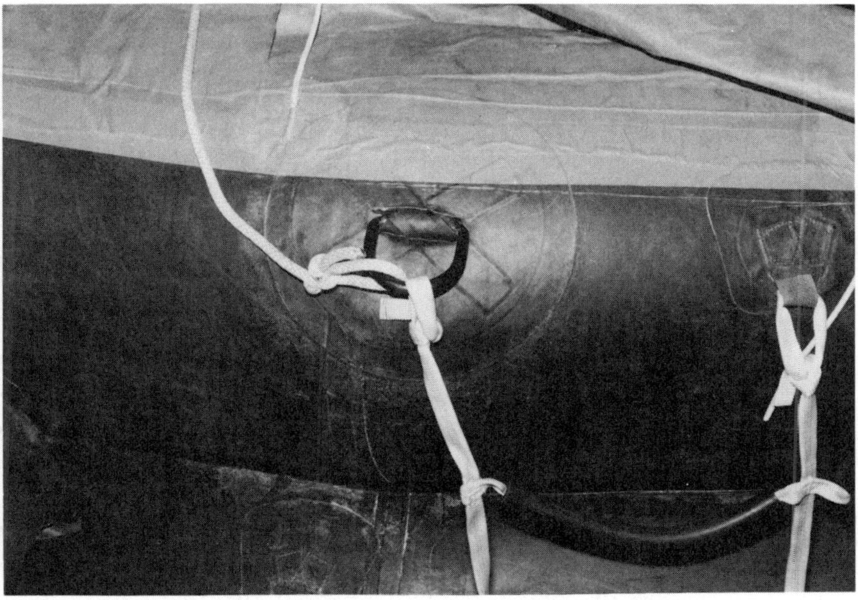

A questionable sea anchor tether on a 25-person life raft. The reasons it might be unreliable are that the tether is fixed to the D-ring with a knot that can loosen; the D-ring patch is placed in tension when the sea anchor is under load and could fail; and the sea anchor is attached by a single tether instead of a bridle that could stabilize the raft's direction relative to the wind. Photo by the author.

tension, and the second is that the single attachment point in the center of the end of the raft would provide little yawing stability to the raft. A preferred attachment is to lead a bridle to both sides of the raft, securing it to patches which are in shear, and thereby enhancing the yawing stability given by the sea anchor.

Smaller life rafts, like the common 8-person octagonal raft used for cruising boats and small commercial vessels, carry a sea anchor with a 12-inch opening, a 3-inch vent, and a 12-inch-long conical body. The drift rate of such a sea anchor-raft combination is about 5 knots in a 60-knot wind.

It is obvious that the 8-person life raft sea anchor (which is really only a drogue) is totally inadequate for the raft from a drift speed standpoint alone. The fact that it is a smaller capacity raft has led manufacturers to assume that a smaller sea anchor is in order. Wrong. The wind drag area of the 8-person raft is about 19 square feet, while the wind drag area of the

25-person raft is 21 square feet—almost the same! In the case of overturning, the 18-foot length and larger passenger weight of the 25-person raft would deter capsizing to a much greater degree than the 7½-foot-diameter octagonal raft carrying a maximum of 8 persons.

One can draw a simple conclusion from the foregoing discussion and that is both the 8-person life raft and the 25-person life raft of approximately the same frontal dimensions should have 2- to 3-foot-diameter sea anchors. Anything smaller invites capsizing in heavy seas and high winds. Along with the appropriate sea anchor size there is the need for a proper attachment of the tether to the life raft. Only a bridle with fittings attached to the sides of the raft so that the rubber patches are in shear should be used. ("Sides" of an octagonal raft are established when the entry flaps of the raft canopy are to leeward.) Because sea anchors and their tethers take such abuse when in use, a spare sea anchor should always be carried in the survival pack.

The shape of the sea anchor is also of importance since it has been pointed out in the U. K. tests, as well as by Shewmon's tests, that simple square canopy sea anchors do not always open of themselves. A properly designed canopy shape will open even without a rigid ring at the opening. Airmen's parachutes have been doing it for years! Be certain that the sea anchor used with your life raft has had successful opening tests run on it.

MAN OVERBOARD POLE SEA ANCHORS

Sea anchors in small sizes are now basic parts of survival equipment for blue water sailing. The familiar man overboard pole on racing and cruising sailboats utilizes a baby sea anchor to slow the drift of the signal pole and life buoy in the wind. A typical assembly of man overboard gear is shown in Figure 21, along with the nominal dimensions of the sea anchor for its use. A good baby sea anchor will have a stainless steel wire spring sewn in the hem of the opening to assure instant opening when it hits the water. Without a sea anchor, the emergency gear could well drift faster than the person in the water and soon be out of reach of his swimming ability.

Some horseshoe buoys are now being fitted with "built-in drogues" to "slow the drift in the water." These are in an embryonic state of development and are inferior to a proper baby sea anchor. The ones I have seen have a visor-type drogue. For the present, the baby sea anchor should be used for man overboard drift control.

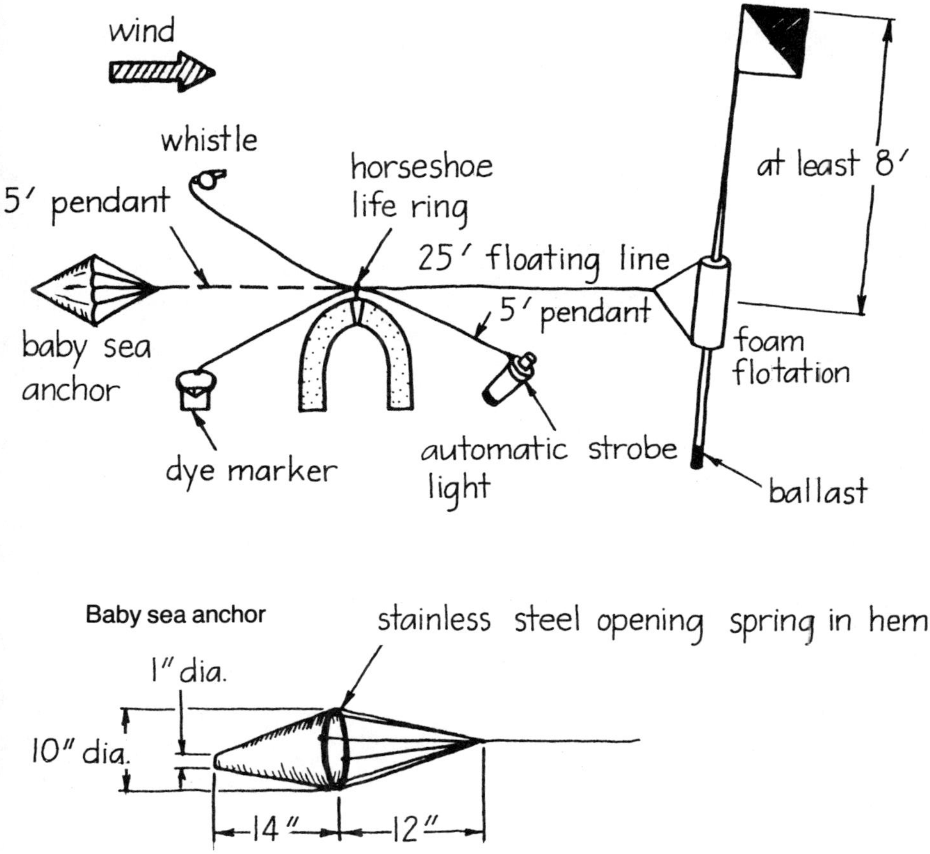

Figure 21. Man overboard gear.

BOAT BRAKES

Boat Brakes is the name of a patented sea anchor of very novel design that has a place on every fishing boat or powerboat up to 26 feet in length; used in tandem, Boat Brakes will also handle larger boats. Basically, the Boat Brake is a parachute that has a trip line attached centrally to the crown, allowing the crown to be drawn towards the opening and thereby reducing the volume and drag of the device. The control rope (trip line) operates independently but parallel to the tether and is virtually foulproof. A system of weights and floats on the canopy opening prevents the canopy from turning and tangling shrouds, tether, and control line.

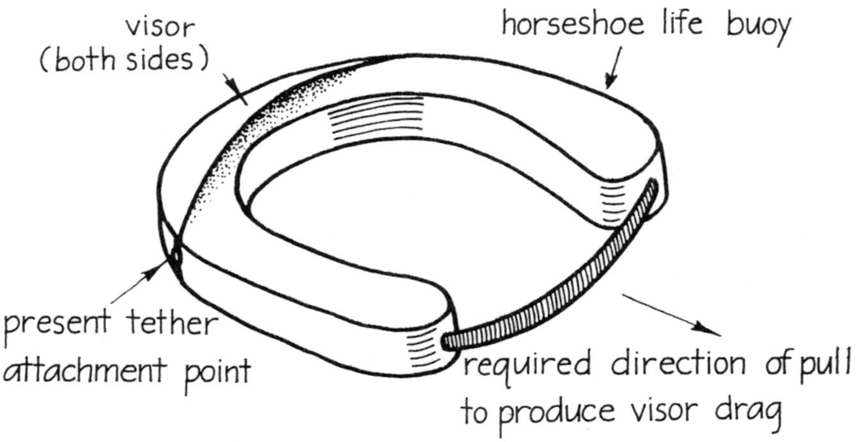

The horseshoe life buoy with a built-in "sea anchor" is the latest offering in safety equipment for the boater. While the idea has merit, the implementation still needs work. The "visor" sea anchor should have its tether attached at the open end of the horseshoe instead of at the closed end as is now the case. The visor sea anchor appears to be a less effective drag device than the baby sea anchor of Figure 21.

The "Captain" series of Boat Brakes has a canopy made of 2.9-ounce fluorescent orange urethane-coated ripstop nylon which is very substantial and durable. The manufacturer recommends the following sizes be used with different boat lengths:

Diameter (feet)	*Volume (gallons)*	*Recommended for*
2	42	Trolling
2½	84	Small fishing boats, canoes, etc., to 14 feet
3	147	Medium fishing boats, bass boats, etc., to 18 feet
3½	236	Large fishing boats, pike boats, etc., to 24 feet
4	351	Charter launches, cruisers, pontoon boats, etc., 28 feet and over*

A "Standard" series of Boat Brakes is also made using lighter (1.8-ounce nylon taffeta) material for use by the casual freshwater fisherman as a drift control device only.

*This includes the use of two Boat Brakes in tandem if required.

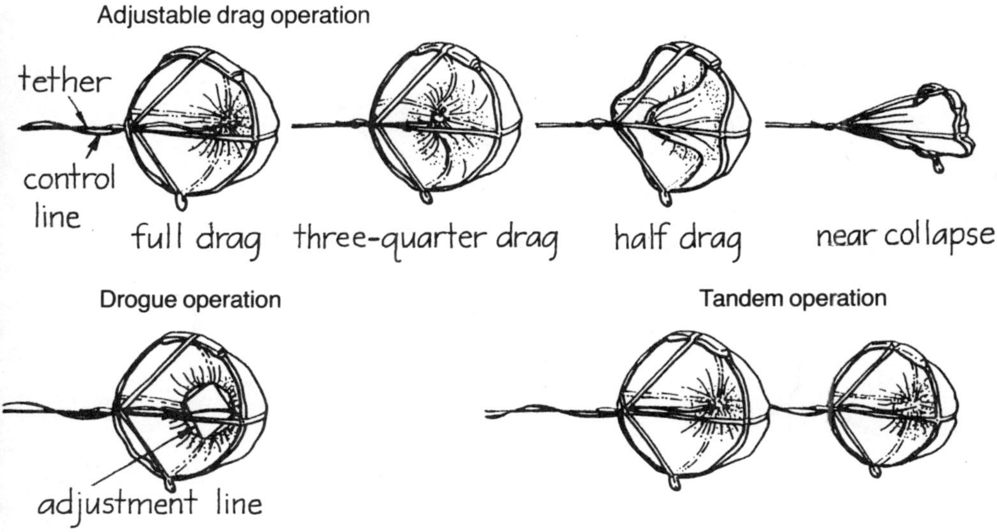

Figure 22. The Boat Brake sea anchor. Courtesy Blue Harbor, Inc.

The Boat Brake comes with one unusual feature for a sea anchor and that is the ability to convert it to a drogue. This is done by partially freeing the adjustment rope (which closes the canopy vent when it is in use as a sea anchor). The adjustment rope can be stopped off with a knot at the desired vent opening to allow the water to flow through the device in a manner similar to a conical drogue. This would certainly have some merit for use on small boats where a single device doing the job of two conserves space and finances.

MENGLER FISHING ANCHOR

From Australia comes a specialty sea anchor designed for deep-sea fishing in windy areas. The Mengler sea anchor is a parachute design that utilizes the umbrella principle for drag control as well as retrieval. It looks like an umbrella when stowed, taking up very little space, and opens like one, forming a parachute canopy for drag.

The secret to the Mengler design is the use of a hollow brass stem to which the canopy shrouds are attached and through which the trip line passes to the float. The length of the trip line between float and canopy crown determines the depth at which the sea anchor will operate. To retrieve the sea anchor, the float is lifted from the water, drawing the

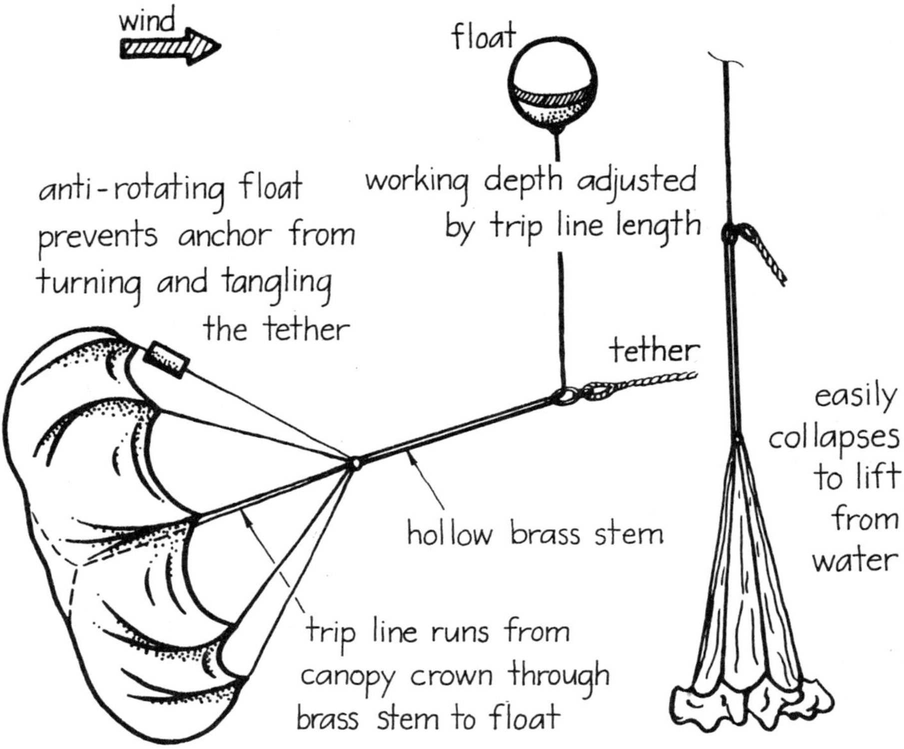

Figure 23. Mengler sea anchor. Courtesy R. D. Mengler.

crown of the canopy inside to the brass stem and essentially closing the canopy. The whole is then lifted from the water and the canopy is wrapped around the stem for temporary stowage until it can be flushed with fresh water, dried, and then stored until the next fishing trip.

The Mengler sea anchor has proven itself in Australian waters for deep-sea fishing by slowing the boat's drift away from the fishing hole. It comes in three sizes: for boats to 20 feet long, 20 to 35 feet, and 35 to 45 feet long.

A fishing sea anchor should not be considered a survival sea anchor because it has neither the size nor the durability to withstand storm conditions.

SIX

Drogue Design

One of the better ways to combat high winds and seas while underway is to tow some drag device from the stern of the boat to slow it down and let the waves pass beneath the keel. Unlike using a sea anchor tethered to the bow of the boat, you only want to slow the boat down, not stop it. The philosophy of the drogue is to prevent the boat from surfing down the front of the wave and also to hold the stern down so there is less chance of pitchpoling. Another factor identified years back by the redoubtable Captain Voss is that you do not want the vessel to turn sideways on the crest of the wave, a movement called broaching, which tends to roll the unfortunate boat over. This is of particular concern when running across channel entrance bars which have waves breaking over them.

The proper speed for a boat being restrained by a drogue is slow enough so that it doesn't surf, but fast enough so that a rolling graybeard does not impact it at full strength. You will also want to maintain good rudder control to assure that the stern is held into the waves even though they may change their approach angle from time to time. Some boats have even been able to steer a moderate course across the waves, making better speed towards their destinations. That, however, must be experimented with so that you don't get the waves too far on the quarter and promote a broach.

So what is a good speed to ride the storm wave? That depends on boat, crew, and wave structure. A racing crew in a boat with a powerful rudder can surf to their heart's content until, possibly, a rogue wave comes along. Cruising boats should probably be held to under 5 knots but not less than 3 knots, the speed at which rudder power becomes seriously diminished. Each boat will behave differently under a drogue and each storm will produce a different set of wave patterns which must be experimented with to find the best (safest) speed.

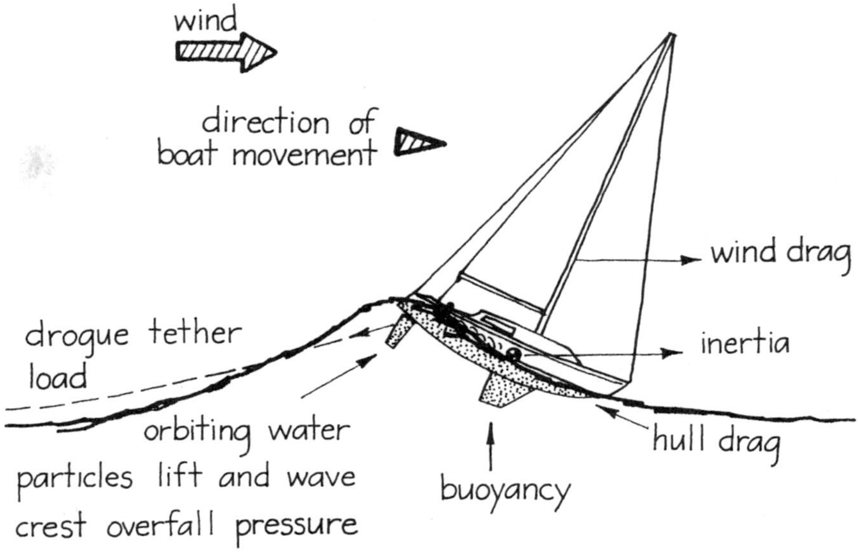

Figure 24. Forces on a boat towing a drogue in heavy seas.

It must be fully understood, when you slow a boat down below the speed of the advancing sea, that an occasional wave can roll aboard, deluging the stern of your boat. You must be prepared with closed cabin hatches and entries and a clean cockpit that you can easily pump out when it becomes a bathtub. While some sailors have negative thoughts of center cockpit designs, they will think again when running before heavy seas.

The design of the stern of the boat will be a major factor in the success of using a drogue under survival conditions. A transom stern shape has the least structural strength while the canoe stern has the greatest potential strength. A transom stern with full bilges will lift easily with the wave while the canoe stern will tend to be buried. A transom stern will take the full impact of the breaking wave while the canoe stern will more smoothly part the crest and let it pass along the sides. Taking the breaking waves slightly on the quarter can materially decrease the impact loads on square transoms.

JURY-RIGGED DROGUES

There are no brakes built into boats to slow them down, so external devices must be used. To date the drag devices employed have been limited only by the imagination of the skippers. Generally they have been jury rigs ranging from automobile tires to ground anchors. The only thing that hasn't been reported is the galley sink, and I would expect that has been tried aboard some boat later posted missing.

Warps are the most commonly used jury drag device and their performance is a greatly disputed issue. Robin Knox-Johnston (*Cruising World*, April 1986) described a system he used when transiting the Southern Ocean. From the stern of his 10-ton canoe-sterned boat he streamed 600 feet of "2-inch"* rope in a large bight (loop) with a smaller bight of the rope at the center of the larger bight. He related that the bight kept the stern into the waves at all times and removed the danger of rushing down the front of a wave and broaching.

David Lewis encountered a 45-knot gale in *Cardinal Vertue* and attempted to weather it by streaming a 37-meter (20-fathom) rope in a bight astern. As he tried to steer before the wind, the yacht became unmanageable, persistently running across the seas and repeatedly broaching-to. The rope hindered steering and each large sea swept it alongside (Lewis 1961).

The larger 9.5-meter (31-foot) sloop *Samuel Pepys*, when running in a hurricane-force wind in the North Atlantic under bare poles, streamed 73 meters (240 feet) of 28mm (1⅛-inch) nylon rope in a bight and two separate 23mm (⅞-inch) hemp ropes each of 36 meters (120 feet). These drags reduced her speed to 3 knots and cut a smoother track through the seas stern-on, but an attempt to move out of the path of the storm by bringing the wind on the quarter at once resulted in heavy water coming aboard. The seas were estimated to exceed 30 feet in height with a length of 400 feet (Hiscock 1981).

Bernard Moitessier, the indomitable French solo sailor, learned the hard way about losing steering control when towing drag devices. On his way to Cape Horn in the 40-foot double-ended ketch *Joshua*, he was overtaken by a great gale and ran before it with five trailing warps weighted with seven 40-pound iron pigs. These, he said, had no apparent effect on the yacht's speed and they prevented *Joshua* from correctly answering her helm.

Since Moitessier didn't like what was happening, he finally cut the drags free and, under bare poles, flew before the gale, steering every inch

*It is believed that the 2 inches referred to is the circumference, a past method of measuring rope sizes. In today's practice it would be a 16mm or ⅝-inch-diameter rope.

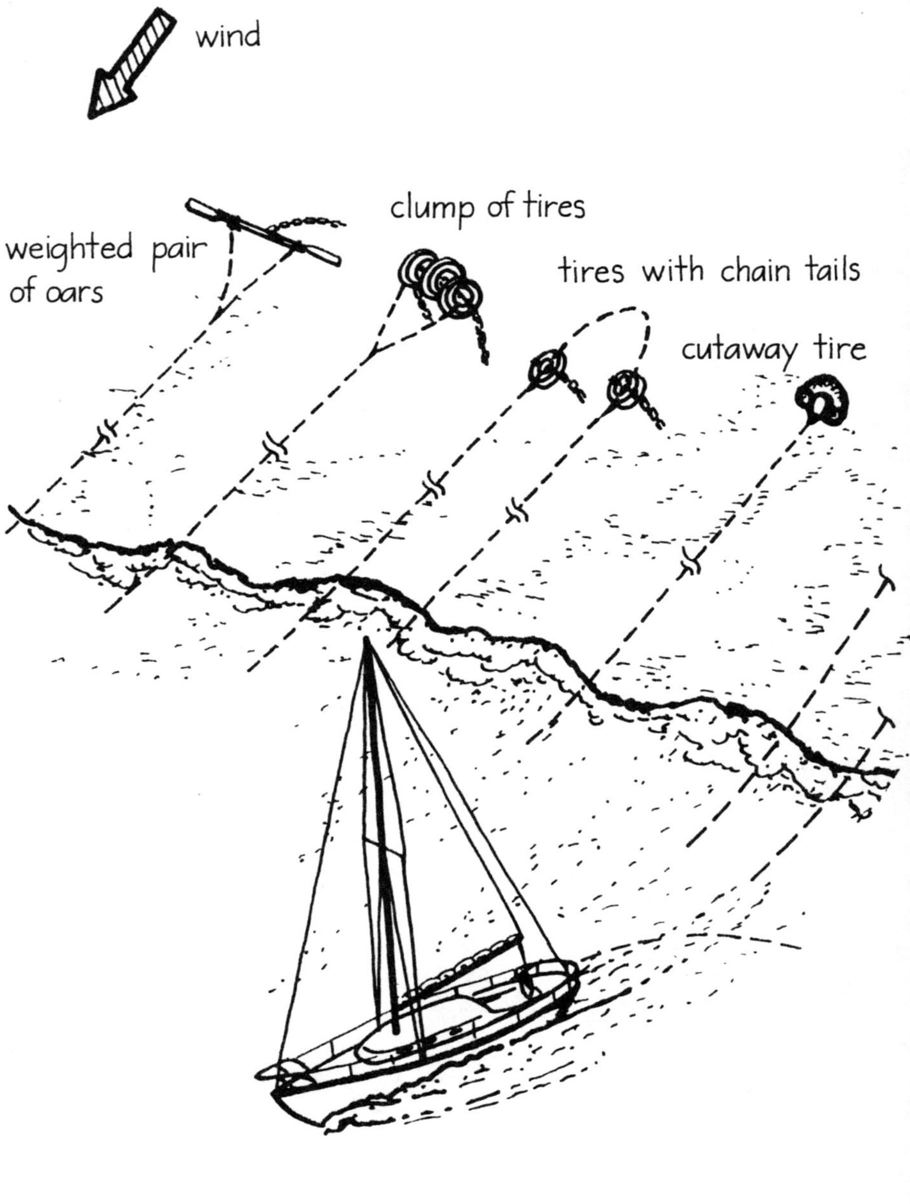

Figure 25. Makeshift drogues.

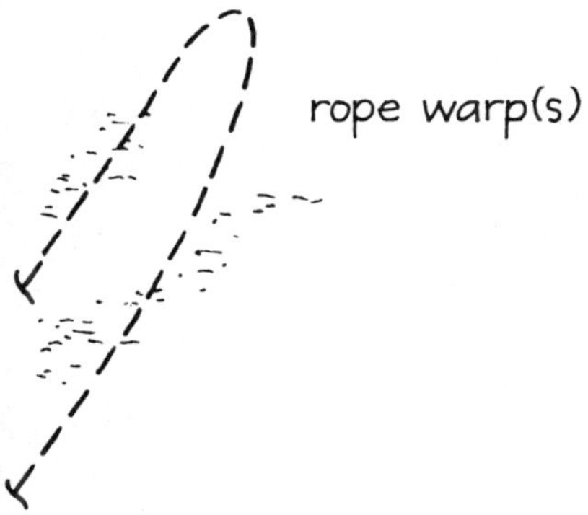
rope warp(s)

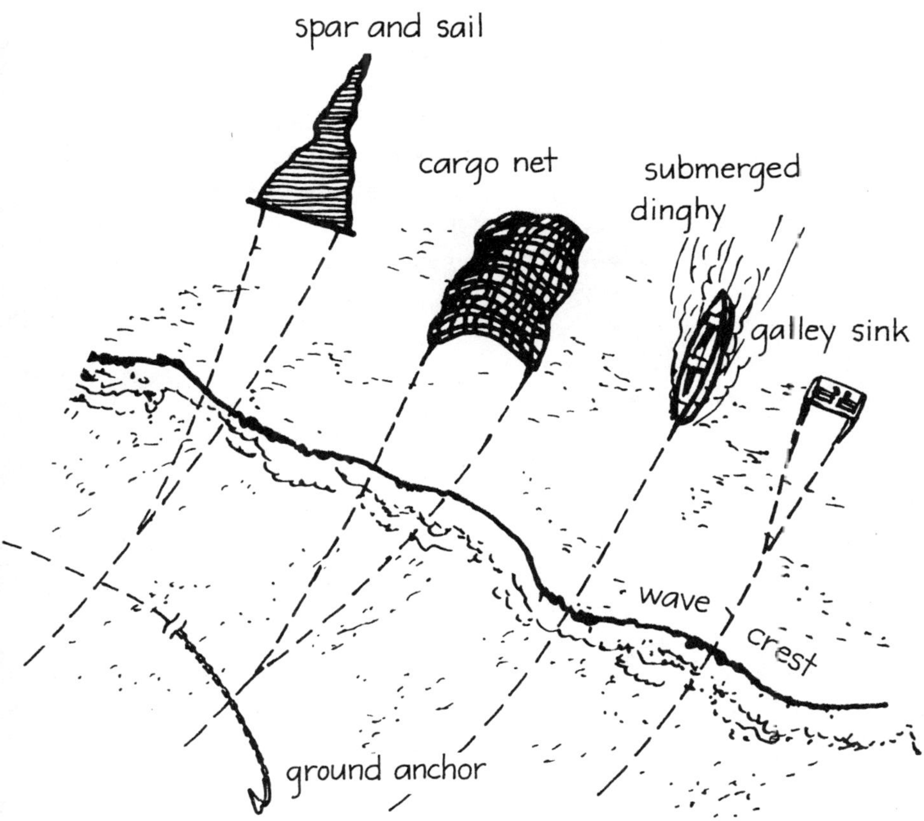
spar and sail
cargo net
submerged dinghy
galley sink
wave crest
ground anchor

of the way and taking each crest at an angle as near as he could manage to 15° on the stern.

Joshua's failure to answer her helm was in all probability due to the warps being attached well behind the rudder. But if the warps had had enough drag to negate rudder control, then they also must have had enough drag to slow *Joshua* somewhat (Moitessier 1969).

Trailed warps just do not have enough drag to be effective. John Rousmaniere in *Fastnet, Force 10* relates how *Grimalkin*, a 30-foot sloop, was trailing 600 feet of line and "she surfed wildly down the face of the waves, like an elevator cut loose from its cable, and threatened to pitchpole, or somersault over her bow. She once accelerated to over 12 knots and tilting forward until she was almost vertical, plunged down the face of a huge wave." This storm not only produced large waves, but they were steep and confused. *Grimalkin* was later ravaged by capsizes and knockdowns until she had to be abandoned.

It is still fashionable in gamming sessions to discuss the towing of warps and one is led to believe that simply dragging a rope from the stern will do the trick. Shewmon, Inc., streamed 170 feet of new 2¼-inch-diameter 3-strand polyester rope behind a tug at 10½ knots. Although the rope had considerable tension in it, the pull was actually less than the 300-pound static friction breakout load of the measuring device. Certainly this adds credence to the inability of simple towed warps (especially without a bight) to be dependable in an emergency.

Ross Norgrove (1980) used a wooden "sea anchor" built like a Fenger drogue to slow down his *White Squall* in heavy seas. He said it was the ideal weapon for her when laid over the stern. What he didn't like about it was the trouble of handling such an awkward device.

Eric Hiscock, the dean of cruising sailors, used a "small sea anchor" to reduce the speed of *Wanderer III* from 5½ knots to 1½ knots while running under bare poles. This allowed him to steer across the wind adequately to avoid an island in his downwind path. On another occasion he lay for four days with his "sea anchor" over the stern and the helm unattended, which is a rare success (Hiscock 1956).

Bob Griffith, while circumnavigating Antarctica in his 52-foot *Awhanee*, found it best in winds of 80 knots to tow a lot of drags, including the spinnaker boom and several anchors and chains. He kept the seas on the quarter all the time because that helped to reduce speed and rolling. This is similar to Moitessier's experience running without the drag devices (Griffith 1979).

One of the most unusual drag devices to be employed as a drogue was on the 35-foot Piver trimaran *Zulane* on a passage from the Cape of Good Hope to the West Indies. Skipper Frik Potgieter deployed his mizzen boom with the mizzen sail still attached and ran off under bare poles. He stated that the sail had a "calming effect on the seas"(Henderson 1979).

Towing tires has been another technique prescribed for holding the speed of a boat down to manageable proportions. I suppose that some boats carry tires as standard equipment to be used as fenders against docks or other boats in rafting, but I see no other use for them. Experience has shown that tires tend to bounce and skip along the water unless they are held down with some kind of a weight. Chain tails seem to work well for single or clumped tires.

Among other things, John and Joan Casanova carried two tires aboard their *Tortuga Too* and tried them as drogues several times, but they found the jerky motion of the boat was too uncomfortable. Then they invented the Easter basket tire illustrated in Figure 26. This cutaway tire gave plenty of drag but produced no lift to make it bounce. For obvious reasons avoid steel-belted or radial tires for this modification.

The number and size of tires to be towed has never been addressed and I would dismiss the whole idea unless I just happened to have tires aboard for some reason that escapes me at this writing. If the tire drogue works, as some say it does, it has to be a very awkward solution in this day of advanced technology.

Moitessier mentioned towing warps with iron pigs at the free ends. Presumably this was to make an arc in each warp to add drag and also to prevent the warps from being cast forward on a breaking wave. Obviously he wasn't too pleased with his idea for he cut them loose and went surfing instead. But the basic idea has been used by others with only a slight modification. Instead of the iron pigs attached to the warps, ground anchors were used—most likely not lightweight, stock-stabilized, or any other design that tends to kite when towed. It would be hard to predict the performance of such a drag device and it certainly sounds like another makeshift jury rig at best.

You have read and heard of successes and failures in the use of jury-rigged drag devices like the foregoing, which may lead you to believe that there is no common thread of application to them. That is not the case. It is clear from the stories that drag devices towed astern when running down large seas are successful *provided that they have enough drag* and that they are attached to the boat in such a manner that *the rudder*

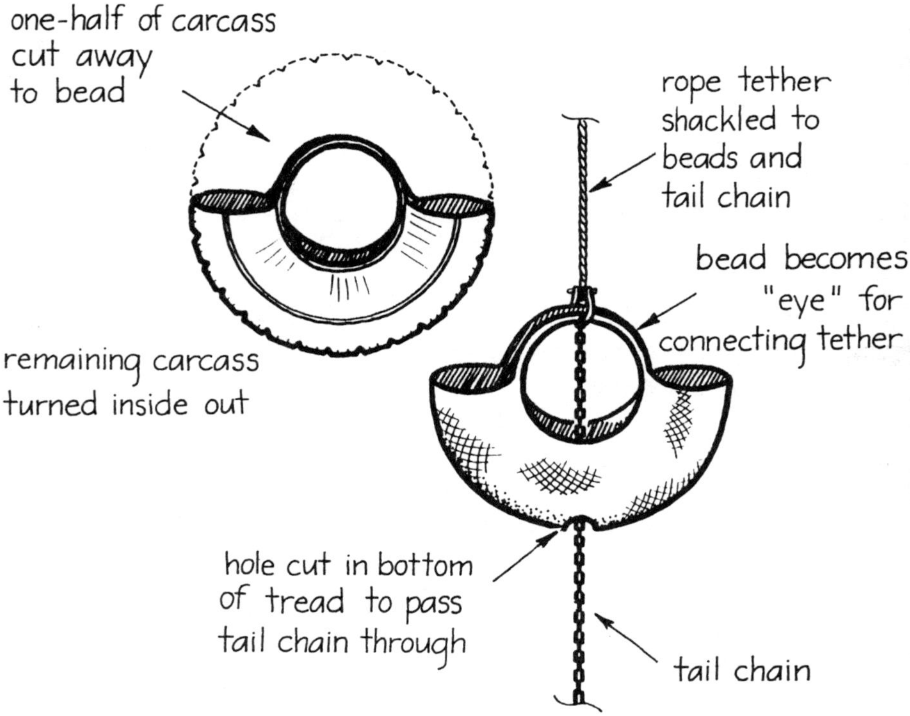

Figure 26. A sophisticated tire drogue.

remains effective. Unless these criteria are met, it is probably more dangerous to tow the jury-rigged devices than not to have any at all.

RESEARCHED DROGUE DESIGNS

Design criteria for drogues are even less well structured than those for sea anchors. It has been a cut-and-try process on the part of those who felt they had to have something called a drogue that they could tow behind the boat when running downwind in heavy seas. Only recently have organized efforts been directed towards quantifying the needs for and the designs of drogues. Along these lines, the drogue designs by Shewmon, Inc., and Hathaway, Reiser & Raymond sailmakers are forward-looking concepts and represent today's state of the art.

Donald J. Jordan's model tests (Jordan 1984) are pioneering work into understanding the mechanism of breaking waves and how small boats can use drogues to avoid catastrophic situations under storm conditions.

Although I previously said that boats do not have built-in brakes, that may change. Warren Luhrs, in his 60-foot 1986 BOC Around-the-World racing sloop *Thursday's Child*, has built in a braking system that he describes (*Yacht Racing/Cruising*, July 1986) as "a scoop-type device that will extend from the transom on two trailing bars, which will be allowed to pivot at the transom." The purpose of the brake, according to Lars Bergstrom, its designer, is to slow the boat down on large waves. "When a boat surfs down a wave, the bow digs into the wave in front of it. It's scary. The brake will allow the boat to stay on the waves and continue to surf with enough sail area up, creating a higher average speed, but cutting down on the very top speed."

Estimating drogue size is more nebulous than estimating the size of a sea anchor. With a sea anchor you want to keep the speed low, so having too large a sea anchor is not a problem. But with the drogue, you are seeking a size that will let you sail comfortably downwind and -wave. It will not let you go too fast for your steering ability nor too slow so that overtaking waves pose a threat to your survival. Since a boat will in all probability be carrying only one size of drogue, it has to be a good compromise.

Shewmon Truncated-Cone Drogues

The Shewmon truncated-cone drogue like the Shewmon sea anchor has been extensively developed over recent years and given severe tests to determine its drag performance, strength, and yaw characteristics (Daniel Shewmon, "Drogues, Tires and Warps," *Cruising World*, April 1984). The measured drag performances are shown in Figure 27 for a range of truncated-cone drogue sizes from 2 to 4 feet in diameter. These full-scale drag tests were accomplished by a powerful harbor tug with a pull of 10,000 pounds.

There appeared to be little tendency for the Shewmon drogues to yaw, which is attributed to the through-flow feature of the design, i.e., a vented cone. This appears to be an essential characteristic of any drogue. Venting is a technique used in many aero- and hydrodynamic designs to stabilize the flow behind an object that is creating oscillatory detached flow. Rotation of the drogue is limited by the use of a 3-pound weight sewed along the hem of the canopy opening.

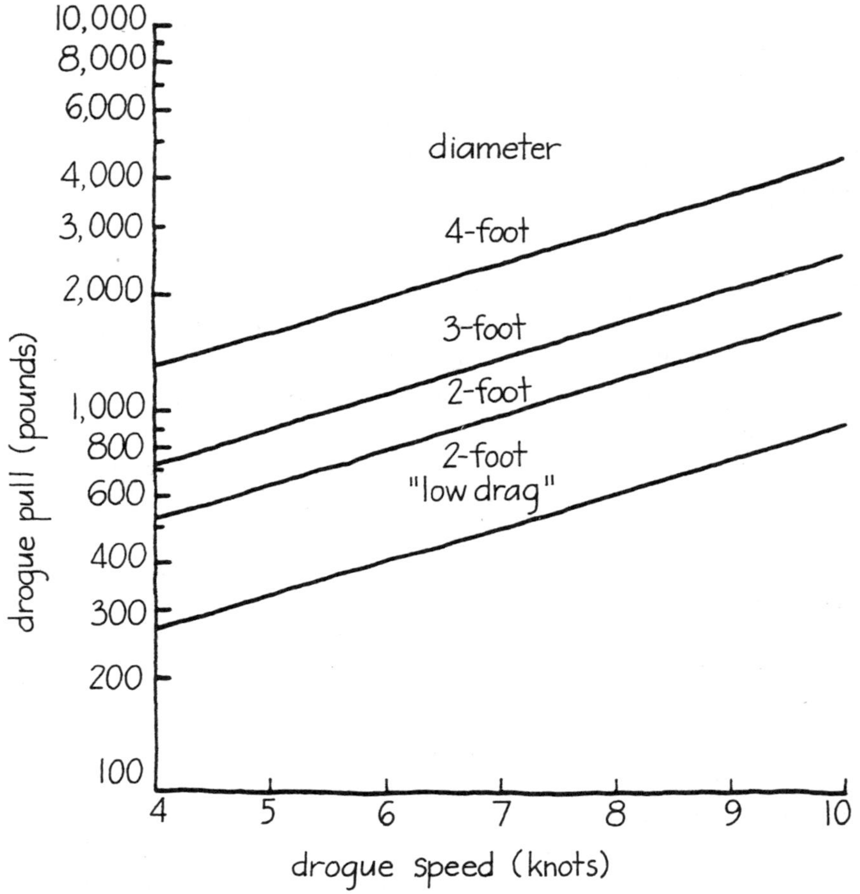

Figure 27. Drag performance of Shewmon truncated-cone drogues. By permission of Shewmon, Inc., © 1986.

Following are recommendations of Shewmon truncated-cone drogue sizes for different size boats:

Boat LOA (feet)	*Drogue diameter (feet)*
30	2
40	3
60	4

Using this table for our example boats, the OI-41 and J/41, you must do some rationalization in your selection since both would appear to use

A 3-foot-diameter Shewmon truncated-cone drogue showing the characteristic large tail opening (vent) of a drogue. In tests this drogue developed over 3,000 pounds of drag at a 7½-knot towing speed. Courtesy Shewmon, Inc.

the same size drogues. The recommended 3-foot drogue would probably be suitable for the OI-41 with its moderate displacement, full keel, and cruising crew. The same size drogue would, however, seem too large for the J/41 with its light displacement, fin keel, and racing crew. For the J/41 a 2-foot-diameter drogue would appear more appropriate. In both instances, if the drag is too great and the boats are moving too slowly for good rudder control, a small (storm) jib should be raised.

Shewmon Variable-Pull Drogues

In 1987 Shewmon came out with a new drogue concept in which the drag can be varied. The drogue body is flat cloth, cut in the shape of a square, hexagon, or decagon depending on its size. The shrouds are attached to the corners of the polygon canopy in a conventional manner but, in addition, there is a trip line attached to the apex of the crown which parallels the tether and runs back to the boat. This trip line can be pulled to invert the canopy from the center, thereby varying the drag of it. In the extreme, the canopy can be turned inside out for retrieval. If the trip line is relaxed, the canopy will refill automatically due to a drag device positioned downstream of the canopy. These drogues are self-opening, seamless, and relatively cheap.

The manufacturer recommends the following sizes:

Maximum boat LOA (feet)	*Drogue size (inches)*	*Maximum drag fully opened (pounds)*	*(at knots)*
30	32	1,700	10½
60	54	5,000	10½
over 60	106	9,000	7

Because of the drag adjustment possible with the Shewmon variable-pull drogue, one size can handle a span of boat sizes.

Galerider Drogues

A unique approach to a drogue is the Galerider, designed and built by Hathaway, Reiser & Raymond of Stamford, Connecticut. This drogue is, essentially, a paraboloid of 2-inch nylon webbing sewn together on 6-inch centers much like an airplane cargo net. The open nature of the design gives it an inherent stability and it would not be expected to oscillate or yaw in use. The mouth is held open by a circle of ⅜-inch 1 × 19 stainless steel wire.

The Galerider was tested by the 20-ton yacht *Southerly* in a passage from New York to Antigua (*Sailor*, December 3, 1985). Winds reached Force 10 as the yacht sailed into the Gulf Stream and seas became confused. First *Southerly* was reefed, then the jib was removed, and later a trysail was set. As things got worse, the trysail was doused and the boat was

Opposite top, The Shewmon variable-pull drogue undergoing a water test. This 54-inch drogue is partially tripped by the trip line to reduce the water drag on it. The float attached to the trip line extension assures refilling after being tripped.

Opposite bottom, The 54-inch Shewmon variable-pull drogue is made in the shape of a hexagon with six shrouds connecting it to the tether. Note the trip line at the center. The manufacturer believes this concept will replace the truncated-cone drogue because of its greater efficiency, variable drag feature, and cheaper and more durable construction. Courtesy Shewmon, Inc.

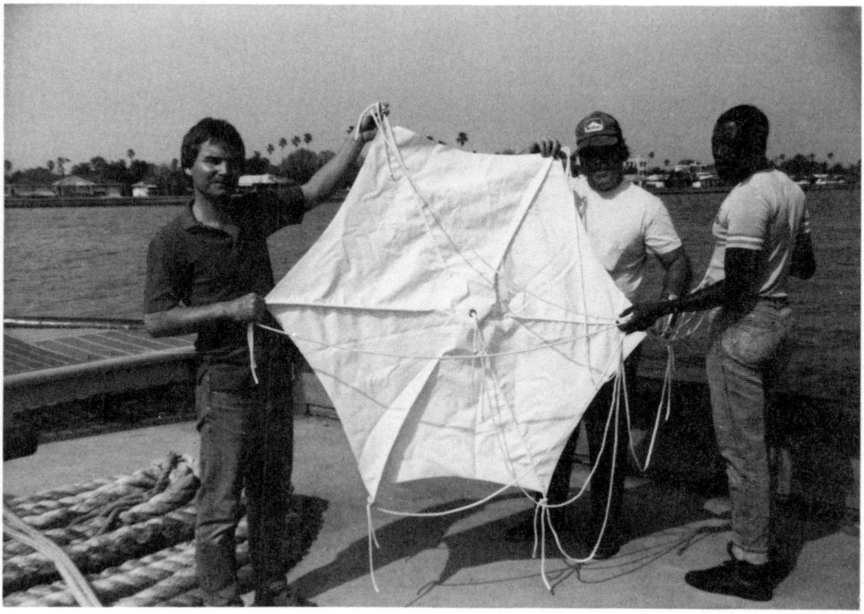

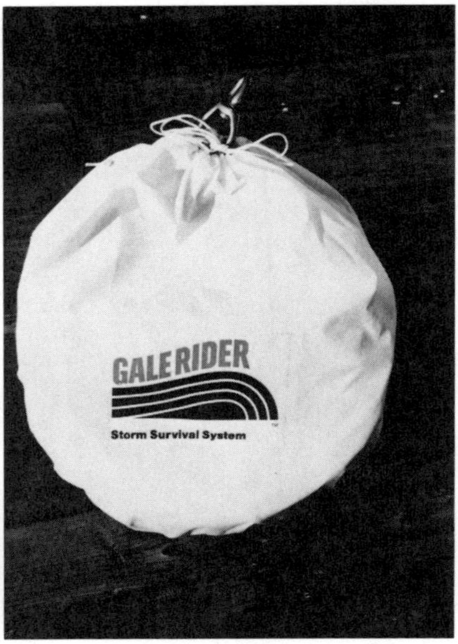

Left, The parabolic shape and web construction of the Galerider are clearly shown here. *Right,* The Galerider drogue packs into a neat 2′ × 1′6″ bag with attaching swivel at the ready. Courtesy Hathaway, Reiser & Raymond, Stamford, Connecticut.

allowed to run under bare poles. Speeds ranged from 3 knots on the backside of waves to more than 12 knots on the face, and broaching became a possibility in spite of the helmsman's best efforts.

A 4-foot-diameter prototype Galerider was deployed on a 1¼-inch nylon anchor rode 200 feet long and belayed to a coffee grinder winch on the afterdeck. (*Southerly* is a center cockpit boat.) The rode was paid out to its full length with no opening shock and the boat soon slowed to a steady 3 knots. As the skipper said, "It went from a charging bull to a docile creature under full control." The helmsman could steer through about 90° and the boat's rolling decreased from full gunnel down on each side to about ±20 degrees. Although several seas boarded the stern, no green water reached the cockpit to threaten the crew. It was estimated that the drogue rode about 30 feet beneath the surface of the water without additional ballast. Finally, it was pointed out that the Galerider drogue was easily winched in with the coffee grinder and lifted aboard by one man.

The very promising Galerider is made in two production sizes: a 36-inch mouth for boats with displacements of 20,000 to 35,000 pounds and a 41-inch mouth for boats with displacements of 35,000 to 55,000 pounds. It is reasonable to consider rigging them in tandem for boats of even greater displacement. In that case, it is recommended that the short tether for the second Galerider be threaded through the aftermost vent hole of the first and attached directly to the shackle holding the first Galerider to the primary tether. The manufacturer recommends that the Galerider should be at least a full sea (wave length) behind the boat to insure that a rogue wave or crest cannot lift both the boat and the Galerider in the same motion.

The smaller Galerider (36-inch mouth) would serve well as a drogue for the example OI-41 cruising boat at 27,000 pounds displacement. It would, however, be too large for the lighter J/41 racing sailboat (15,500 pounds displacement), slowing it down too much for good rudder control. In that case, a larger jib would have to be raised to ensure sufficient speed for active steering. Because there are so many racing sailboats in this light displacement category, a smaller Galerider would probably be very welcome.

It is obvious that one size of drogue will not be perfect for all combinations of wind speed, boat speed, and sea conditions. But, since it is impractical to carry more than one drogue on a boat, the best "average" size should be selected. Even with a drogue, a good measure of seamanship must be employed because each sea is different and only the crew on board using all of its available resources can successfully bring the boat through a storm. You cannot depend solely on mechanical devices to solve all the problems that the ocean can throw at you.

Jordan's Drogue Research

Donald Jordan has been doing a lot of innovative research on the use of drogues to improve the safety of sailing yachts and life rafts (Jordan 1984, and see *Sail*, June 1986, September 1985, December 1982, and February 1982; and *Multihull*, January/February 1987). His tests were conducted with scale models ballasted to give the proper dynamic characteristics. Breaking waves were simulated with horizontal jets of water, and by the wake of a towed boat.

Jordan's work is still continuing, but he has already identified a number of tentative design and operating considerations pertinent to the use of drogues on boats in survival storms. In summary these are:

Modern sailing yachts are capable of lying ahull in large *non-breaking waves* with no significant risk of capsizing.

A drogue whose diameter is about 10 percent of the boat's length is adequate to hold the stern of a *non-steered boat* into the wind in *non-breaking waves* and will prevent it from surfing down the face of the waves.

A drogue diameter of more than 10 to 15 percent of the boat's length is required to stabilize it in *breaking waves* when there is *no active steering*.

If the boat's stern is held *directly into the breaking wave* by *active steering*, a smaller drogue will suffice than if the stern is at a broader angle to the wave. (This is sufficient reason to maintain active steering while running before the seas either with or without a drogue.)

In a *breaking wave strike*, the drogue must act quickly and powerfully to bring the stern of the boat into the fast-moving water of the breaking wave. If the drogue is too small, or the tether is too elastic or has too much slack in it, the boat may broach and capsize.

Wind does not appear to be a significant factor in the capsizing of a boat. With all sail off, there is little force due to wind pressure, especially when you consider that the boat is partially shielded from the wind by the wave itself.

This author would caution the reader to be wary of using a high-drag drogue whose diameter is as large as 10 to 15 percent of the boat's length without considerable experimentation in less than gale force winds. Jordan's work assumes non-steered boats at 45° to the breaking waves. It becomes the drogue's job to slue the boat downwave and hold the stern into the break. This is a very demanding assignment and produces a severe loading on the transom as the wave strikes. Broad, flat transoms (as in an OI-41) are particularly vulnerable. Low, duck-tailed transoms common to modern racing boats (such as the J/41) would be totally inundated, putting crew members who are in the cockpit at extreme risk.

If the boat can be actively steered downwind and capable helmsmen are available for this, then a drogue under 10 percent LOA would appear to be quite effective.

As a result of his research, Donald Jordon has conceived a prototype drogue design suitable for a 30-foot fin keel sailboat (*Sail,* June 1986). His concept, with approximate dimensions, is illustrated in Figure 28. The whole idea of the "series drogue" is to maintain a more or less constant pull by the drogue (a) to prevent the boat from having time to turn sideways to the waves and (b) to prevent the drogue and tether from possibly being cast forward by a breaking wave and becoming tangled and useless. The drogue is weighted to make it run deep below wave action and to maintain a small tension in the tether at all times.*

The series drogue is a fresh approach to solving the problem and should overcome most, if not all, of the problems of drogue use. Consideration must still be given to whether you intend to steer the boat downwind or to let it be completely restrained and controlled by the action of the drogue. If you want to steer the boat when running under a drogue, and are capable of it, then you would not want the bridle attached at the quarters that would hamper steering (see Chapter 7). On the other hand, if you want the drogue to take full control of the boat, then you had better be certain that the boat is able to take the full impact of breaking waves on the transom and in the cockpit. Jordan recommends that the crew take refuge belowdecks, using safety harnesses akin to seat belts to hold crew members safely in their berths.

This is not a scheme to be tried with just any boat since today's boats are structurally unsound for such loadings. Transoms must be heavily reinforced; the cabin structure must be strengthened to resist tons of blue water that would descend on it in a breaking wave (center cockpit boats would have great appeal in such a situation); the cockpit tub must be made smaller and provided with larger drains to assure that the water will quickly run out and not drag the stern down into the next wave; and the bridle attachment bolts would have to be backed up with straps to spread the loads through the hull structure much like standing rigging chain plates.

Jordan's work is still in progress and he points out that the design shown in Figure 28 is not a specification but rather an illustration of the magnitude of the design problem.

---

*Jordan's article in *Sail* indicates the use of a stock-stabilized pivoting fluke anchor of 25 pounds or more. This anchor style is known to "kite" when dragged through the water and could induce some unwanted motions at the end of the drogue. I would suggest the use of a non-kiting anchor or a lead weight.

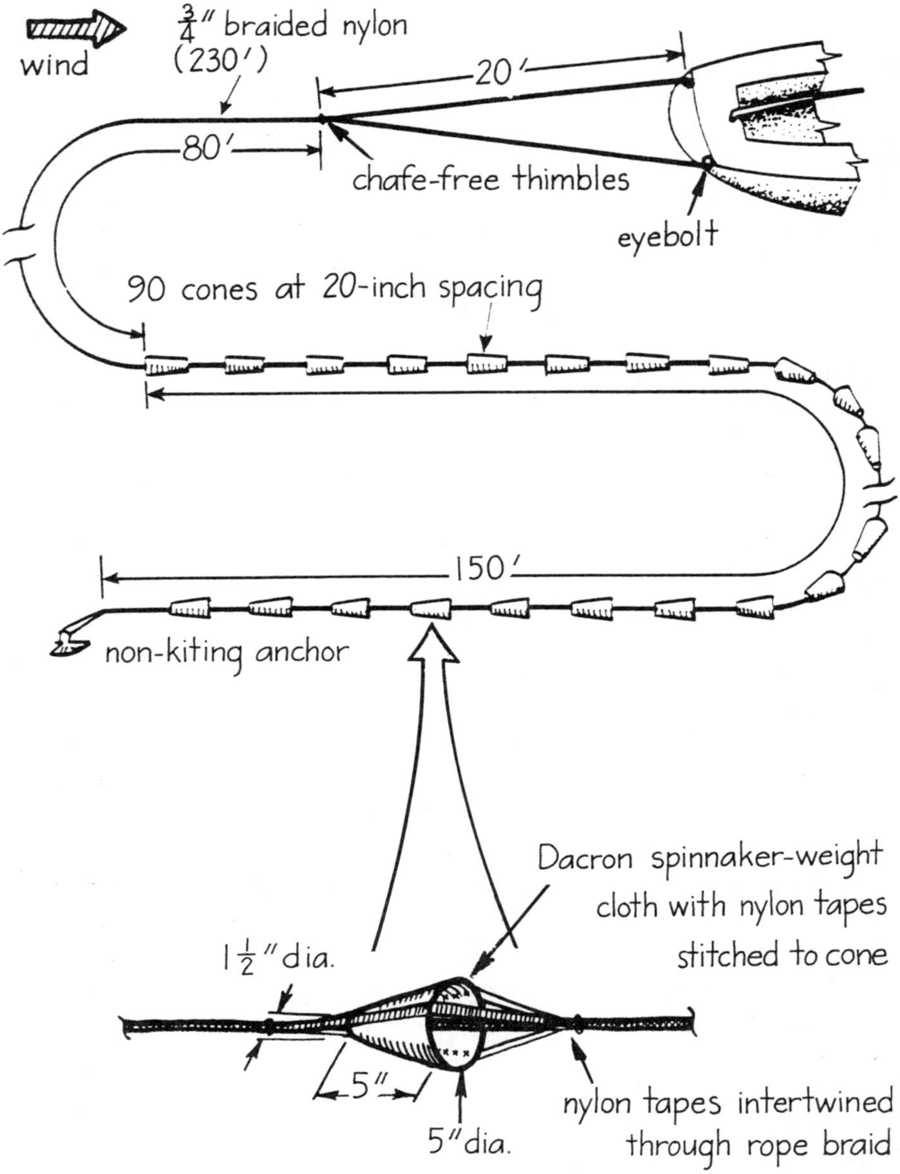

Figure 28. Series drogue by Jordan, sized for a 30-foot LOA fin keel sailboat.

MAKING YOUR OWN DROGUE

Making your own drogue is certainly within the capability of canvas craftsmen. The techniques are similar to heavy sail making. K. Adlard Coles in *Heavy Weather Sailing* had something to say about the use and construction of drogues. While in a Force 11 gale in the approaches to the English Channel, Coles deployed a "small sea anchor" from his 32-foot sloop *Cohoe:*

> First we experimented with the anchor in the conventional position forward. Here it did no good. The yacht continued to lie broadside to the seas and the anchor lay away on the windward quarter. So we passed the nylon warp aft and made it fast. At the stern it had much the same result as is achieved by towing warps. It seemed to have a steadying effect, and the yacht lay with the seas on her quarter.

The sea anchor used as a drogue was later lost due to a failure of the ring and Coles adds:

> The loss of the sea anchor [*sic*] confirms the experience of other deep-sea sailing men that tremendous strains are imposed on a sea anchor and its gear, which should be very strong.

Van Dorn in his book *Oceanography and Seamanship* gave the basic design and dimensions for a make-it-youself drogue. These are reproduced in Figure 29. Particular attention should be paid to strong stitching and quality materials because a drogue takes quite a beating in 24 hours of use, as both Hiscock and Coles noted.

Van Dorn also addressed the size question. He suggested that a truncated conical drogue have a diameter of 1/17th the length of the waterline. For the OI-41 cruising sailboat (34 feet LWL), that yields a 2-foot-diameter drogue and for the J/41 racing sailboat (36 feet LWL), it yields a 2.1-foot-diameter drogue. The 2-foot drogue on the OI-41 is, perhaps, small and you might have to lower all sails to keep the boat speed down to a manageable level. On the other hand, the drogue drag may be too great for the J/41. Additional headsail could then be added to keep the boat speed high enough for active steering and to prevent the overtaking waves from having an unnecessarily severe impact on the transom.

Two other sources of make-it-yourself drogue instructions are contained in *Modern Sailmaking* by Percy Blandford (1978) and *The Finely Fitted Yacht* by Ferenc Mate (1979). In both instances the authors refer to their products as sea anchors when in effect they are too small for sea anchor use, but would make suitable drogues.

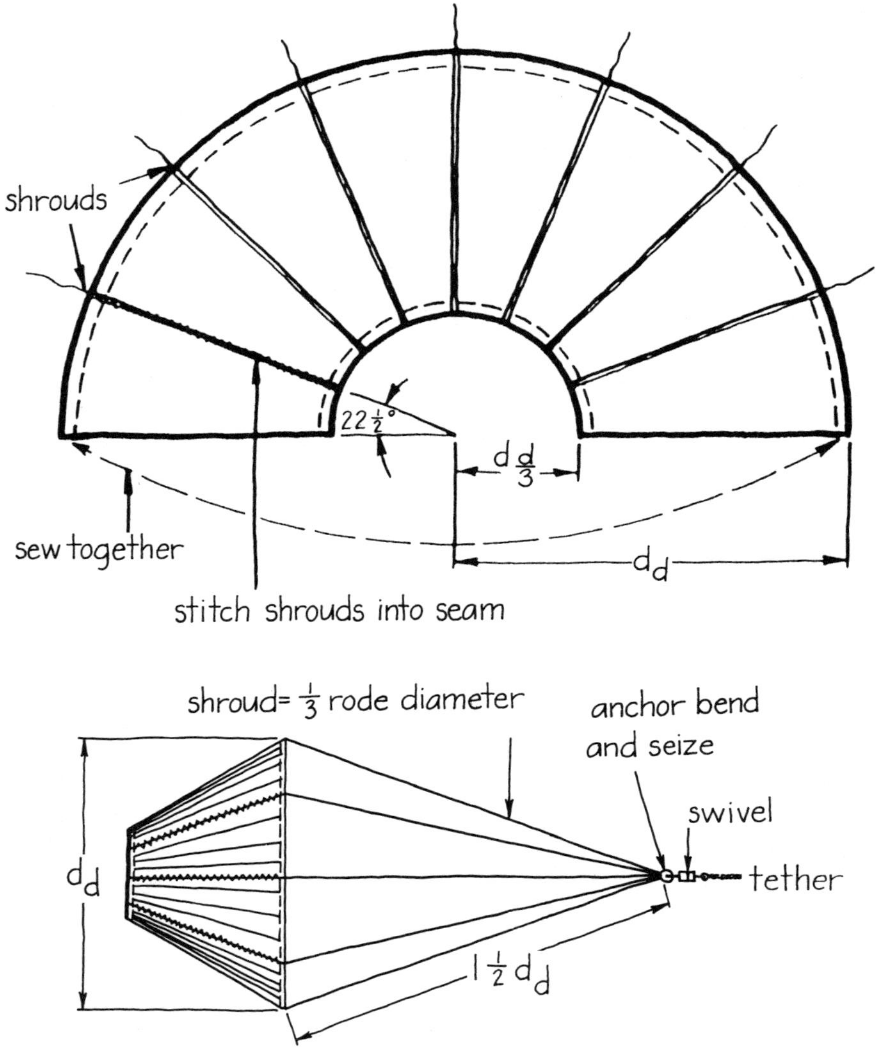

Figure 29. A do-it-yourself drogue design. Reprinted, by permission of the publisher, from William G. Van Dorn, *Oceanography and Seamanship*, Dodd, Mead & Company (New York, 1974).

SEVEN

Use of Drogues

Running before a sea is a tactic used by many boats that are structurally sound and fully controllable. It is an exciting and exhilarating ride, but very demanding on helmsman and rudder. Running with too much speed, though, can be just as dangerous as running with too little speed. Loads on the rudder (and helm) increase as the square of the boat speed; for example, a speed of 15 knots produces forces 2¼ times as great as those at 10 knots. Few boats are built to withstand such forces for long periods of time and few helmsmen can long endure the intense concentration required.

It is the helmsman's job to steer before the waves running off at a slight angle (Moitessier recommended 15°, others suggest as high as 30°) when a breaking crest is about to overtake the boat. Generally you never steer dead downwave because of the risk of the wave breaking and pitchpoling the boat into the trough. Boats with fine bows, unable to lift, are particularly susceptible to this embarrassing predicament.

Short of pitchpoling, an overtaking breaker may "poop" the boat with seawater by the ton filling the cockpit and finding its way below through loose binboards, ventilators, broken skylights, etc. The added weight of water and its fluid nature may render the boat marginally controllable and it could take up a position parallel to the waves. There it will be at risk of rolling over if another breaking crest reaches it before the trapped water can be emptied, the boat accelerated, and rudder control reestablished.

Should the helmsman steer the boat too broad on the wave, there is the likelihood that a breaking wave will drive the boat into a full broach, dumping tons of seawater onto it while simultaneously rolling it over. Neither the pitchpole nor the broach is an attractive alternative to maintaining full control of the boat running downwind. But there is a useful means to bolster downwind control and that is with a drogue.

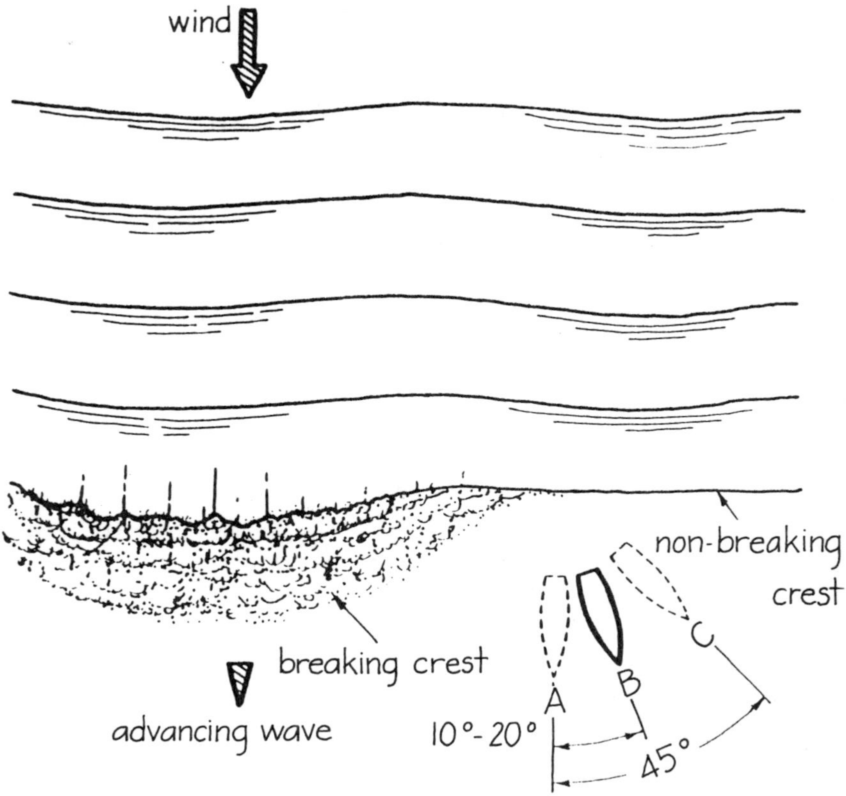

Boat A is liable to pitchpole.
Boat C is liable to broach.
Boat B has the best chance at steering survival.

Figure 30. Steering angle down a breaking wave.

Drogues are intended to slow the forward speed of a boat and to help keep the stern of the boat directed into the oncoming waves. In gales where there is plenty of sea room, it is usually desirable to keep a functioning boat moving and under control, but not to get it in such a position that it will surf down the face of the waves or broach. To avoid broaching, the stern of the boat must be kept 10° to 20° to the waves, for if the boat turns very far off the wave, the crest can slue the stern around, putting the hull broadside to the wave, with the danger of rolling it over. This must be avoided at all costs.

Depending on the wind's speed, you may or may not want to set a headsail—say a Number 4 jib, a storm jib, or a staysail. The idea of a small headsail is to help hold the bow downwind and add some drive for better control. If you have more speed than you can handle just running under bare poles, forget the headsail. But if you do raise one, raise it as high as possible in order that it can see the wind when the boat is in a trough.

The preferred speed of the boat depends on what the boat and the helmsman can handle. A moderate to heavy displacement cruising boat may ride best at between 5 and 10 knots while a light displacement racing sailboat with a capable helmsman may have a safe, controllable sleigh ride at 10 to 15 knots or more. If the waves aren't breaking too severely, go for it. If the waves are steep, crisscross, and break frequently, temper your speed to what the helmsman can safely handle, shorten sail, and run before the wind with bare poles and drogue streamed.

The need for adequate speed when running before the seas was demonstrated by the 42-foot light displacement sailboat *Police Car* in the 1979 Fastnet Race. While running before the gale the crew thought that a speed of 5 or 6 knots was safe until *Police Car* experienced three 120° knockdowns. That was enough to convince them to increase their speed to about 8 knots by changing the trim of their jib. Now the helmsman had enough maneuverability to avoid breaking crests and the boat was no longer victimized by the huge seas (Rousmaniere 1980).

You should note that although a wave may be hundreds of yards or more in length, it will probably be breaking in only one or two places at any one time. Those breaks are what you want to avoid. Take a tip from the board surfer—ride the crest to one side of the break but, unlike the surfer, do not hold your position on the forward face of the wave. Let it pass under you, for if it should break, you will be submerged, just like him, but with poorer chances for recovery.

The idea of the drogue is to slowly pull the boat backwards relative to the wave so that it will pass under the boat. Of course, not all boats will react the same to a wave passing under the stern, especially if it is a breaking wave. Broad transoms are most difficult to hold into the wave and long counter sailboats are next in difficulty. Least difficult are double-ended boats, with canoe sterns being the best of all. But good, better, or best, the problem is really one of being able to steer the boat for long periods of time and to stay alert to the continuous onslaught of overtaking graybeards. Capable helmsmen on racing sailboats are good at this until the point where they become tired, then one miscalculation can spell doom.

John Rousmaniere in his morbidly exciting book *Fastnet, Force 10* describes in vivid detail the harrowing adventures of *Windswept,* an Offshore One Design 34 class that was racing in the 1979 Fastnet Race. Under storm jib alone, with 60 knots of wind, *Windswept* sailed fast and in full control until the mast began shaking violently as the boat was battered by high seas. The skipper ordered the storm jib down and decided to let the boat ride ahull, a sometimes safe and comfortable technique used by many boats for riding out storms. And so it was for *Windswept* until a huge wave rolled her over, from which she fortunately snapped back to her feet.

Realizing that lying ahull was not safe in this maelstrom, the crew improvised a sea anchor out of a sail bag wrapped with 360 feet of line having an anchor affixed to it. This was deployed over the bow on a mooring line. The jury-rigged sea anchor had too little drag to hold *Windswept*'s bow into the wind steadily. *Windswept* again fell off until she was broadside to the seas, when another large wave assaulted her bow and smashed her upside down a second time.

With extensive damage to the boat, an injured crewman, a disabled engine, and a useless life raft, but with the sailing rig still intact, the crew moved the inadequate sea anchor to the stern, making it into what proved to be a lifesaving drogue. The improvised drogue immediately took hold, slowing the boat's downwind rush and making it manageable this time while taking away the threat of a pitchpole.

Windswept's crew tried to improve on the drogue. A 6-gallon jerry can of water provided some drag, but loops of line hung over the stern (warps in a bight) did nothing for the boat. When wind and seas later moderated, the crew took in the improvised but effective drogue, set a Number 2 jib and worked their way to safety at the Crosshaven Yacht Harbor, Cork, Ireland.

Many other boats in the race were in the same predicament as *Windswept* though none, as far as is known, attemped to set a drogue, but simply hove to or laid ahull, tactics which did not work out in the severely confused seas of this gale. We know that 5 boats sank and 19 others had to be abandoned for various reasons. Certainly those that had retained their sailing rigs after the initial rolldown could have had a better chance at survival had they used *Windswept*'s technique.

There is a postscript to the Fastnet disaster provided years later by Bill Burrows, chief engineer for the Royal Navy Lifeboat Institution. He helped bring in three disabled Fastnet boats in a 21-hour period of rescue. In an interview with Peter Spectre (*The Yacht,* April 1987) Burrows said:

Look, you get 300 yachts in poor weather, and you're going to have some in trouble, almost certainly. But the majority of the trouble was hysteria created by the situation and by inexperienced crews. And that it was. They were blaming rudders and such, but none of those rudders would have snapped if they had put drogues out, and storm jibs and run before the weather. They were under bare poles, most of them, and they were getting up on the seas—and the seas were about 45 feet, not what we around here call *big*.

They got up on these seas, and they were running. When the boats were starting to broach, what the helmsmen were doing was hauling on the rudders to stop them from broaching. They were putting too much bloody strain on the rudders, and they *had* to go.

Yes, I know they were racing sailors, not cruising men, but that's no excuse. We went out that night and we passed a little old hooker sort of thing with a family of kids aboard, and she was going away to Ireland with no trouble at all.

TETHER DESIGN

A proper tether is essential to the use of a drogue. Like that for a sea anchor, the drogue tether must be long enough to span one wave length (crest to crest distance), and be strong enough to withstand the drag load of the drogue—including an allowance for surging—and have some elasticity to it. Also, like the sea anchor tether, the rope construction is of importance.

As far as rope construction is concerned, braided rope is by far the best, with proper end fittings to connect to the drogue and to a bridle at the stern of the boat. A bridle is essential because few boats have a sturdy Samson post located forward of the rudder to which the drogue tether can be attached. This feature will be discussed later in detail.

Unlike the easy rigging of a sea anchor tether to deck gear that is already in position for handling the ground anchor off the bow, there usually is no substantial ground tackle handling gear at the stern of a boat. Therefore, it will be necessary to create a tether bridle that can take advantage of cleats and winches which may be symmetrically located along the side decks. A bridle is made to which the tether is attached and the bridle is then secured to these cleats or winches with due consideration for any chafing that may occur. The ropes making up the bridle should each be equal in strength to the drogue's tether since as the drogue yaws, its full drag can be applied to one or the other bridle line.

The wind drag on the boat can be estimated as previously described for the sea anchored boat or, given a stock drogue such as one of the Shewmon drogues, the drag of the drogue can be determined from the manufacturer's test data. Whichever method is used, surge loading is estimated to double the steady load. This total load (wind plus surge) should be no more than 50 percent of the breaking strength of the tether. A larger margin should be allowed if the tether is a piece of rope used for other purposes (such as an anchor rode). A dedicated tether and bridle assembly should be considered a good investment for any boat that regularly operates in the open ocean and must often face gales or worse storms.

Tether elasticity is a rather moot point at this stage in the state of the art. Elasticity is desirable since it reduces the loads on the tether and its attachments to the boat. However, too much elasticity will render the drogue less effective in sluing the stern of the boat into the breaking wave. But then, too, a tether made out of highly elastic nylon (i.e., twisted 3-strand nylon) may have little elasticity in it when subjected to a sharp transient loading. The elasticity of nylon is known to be time-dependent. A slowly applied load can stretch nylon up to 50 percent of its length, whereas a rapidly applied load (as in the sudden impact of a breaking wave against the transom) may not give the nylon sufficient time to develop its elastic qualities. In the latter case the dynamic loading will have to be absorbed by the tether and its attachments.

Kenneth Bitting conducted a test program for the U. S. Coast Guard on the dynamic behavior of double braid and 8-strand plaited nylon and polyester ropes as used in ship mooring applications (Bitting 1980). His work confirmed that the elasticity of synthetic ropes is highly dependent on the rate of load application. Some of his results:

Dynamic elasticity of nylon and polyester ropes may be only 25 to 30 percent of the static elasticity when instantaneously loaded.

Dynamic elasticity of nylon and polyester ropes is markedly decreased after exposure to years of saltwater immersion.

Nylon and polyester ropes lose up to 50 percent of their tensile strength after 4 to 5 years of saltwater immersion.

Nylon rope can withstand higher dynamic loadings if some tension is always kept on the line.

What we do not know is the rate of onset of dynamic loads applied by the sea anchor or drogue in order to quantitatively assess the loss of elasticity in a tether. Since boats using sea anchors have never reported a tether failure due to jerking (abrasion doesn't count), we have to assume for the present that the rate of load application is less than used in Bitting's work and that the rope's elasticity is available.

If you have ever towed another boat, you already know that a tow drastically reduces the maneuverability of your own boat unless you properly locate the towline attachment on your boat. Criteria for this location are shown in Figure 31.

It has been amply demonstrated in tugboat operations over the years that the towline attachment should be made just aft of the longitudinal center of buoyancy in order to preserve maneuverability. The longitudinal center of buoyancy is generally located at a point 54 to 59 percent of the waterline. This puts it pretty far forward on work- and recreational boats where deckhouses and other structures get in the way of an attach-

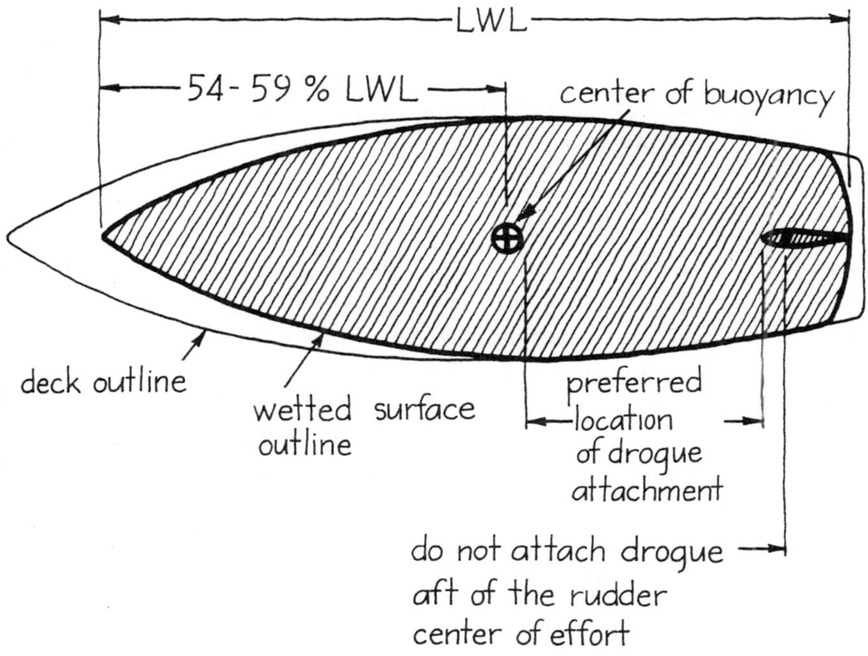

Figure 31. Where to position the drogue attachment.

Tournament ski boaters have found that placing the towing pylon near the center of the boat greatly enhances maneuverability. If the ski line is attached to an inboard boat as far aft as the transom, it will induce rudder stall. These same arguments say that a drogue should be secured well ahead of the rudder in order for the boat to be able to maneuver when running downwind and -wave in heavy weather. Photo by the author.

ment. You may have noticed that tugs keep a clear deck aft so that the towline can be brought forward closer to the center of buoyancy.

In the practical case the tether (or its bridle) should always be forward of the rudder center of effort as shown in Figure 32. To attach a tether aft of the rudder is to cause the tail to wag the dog. Admittedly it will be difficult to attach the drogue tether ahead of the rudder and impossible to attach it at the longitudinal center of buoyancy but all effort must be made to preserve the maneuverability of your boat. When a large graybeard rumbles down on your transom, you will be desperate to square your boat away before it hits, and rudder control is all that will do it.

DROGUE DEPLOYMENT AND RETRIEVAL

Deployment of the drogue takes place over the stern of the boat after the bridle, tether, and drogue are fully connected to each other

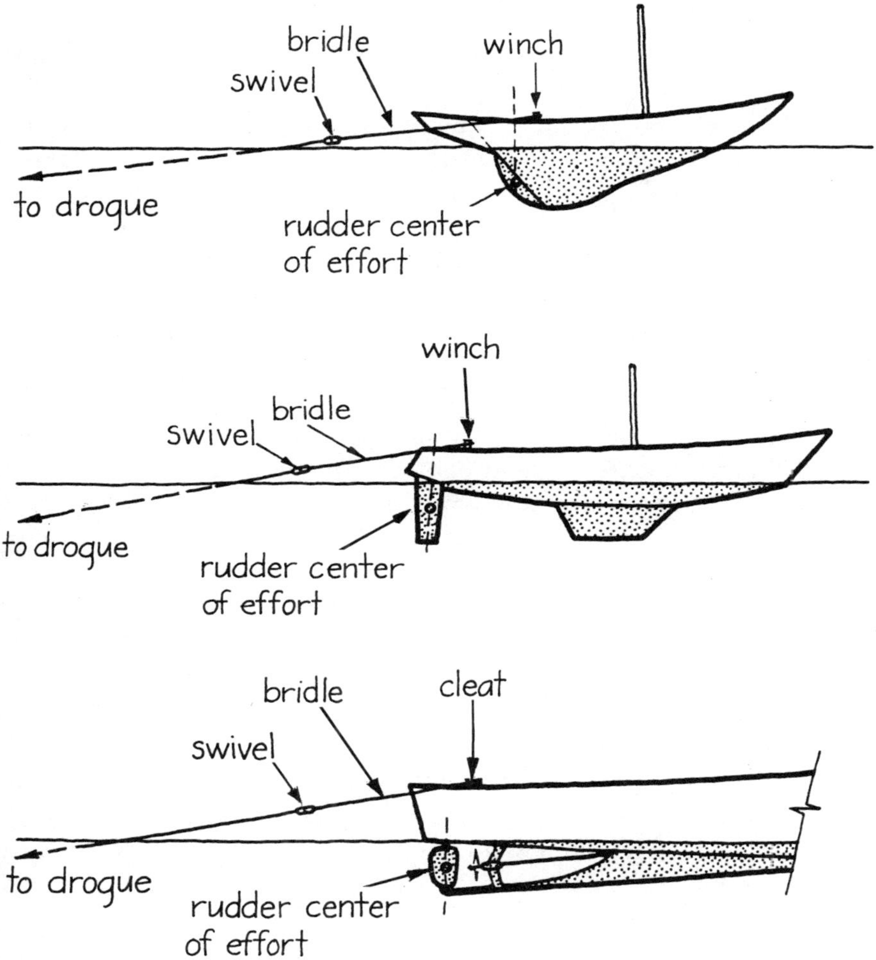

Figure 32. Typical drogue bridle attachment points.

and to the boat. The assembly should be carefully checked for proper knots and splices, safety-wired shackles, and freedom from entanglement with itself or any part of the boat and crew. The tether is deployed starting with the end near the bridle. This is let over the stern so that the drag of the water starts to pull it along in a bight. More and more of the tether is eased over the transom until the drogue, itself, is ready to go. It is then stretched out and cast overboard away from the tether so there is no possibility of it getting tangled.

The drag of the bight in the tether may produce upwards of 100-200 pounds loading. While a strong crewman could handle it with gloved hands, it is advisable to snub it up on a winch, Samson post, bitt, or cleat for safety.

There will be an almost unnoticeable jerk when the drogue has straightened out the tether to its full length, the reason being that the snaking line in the water absorbs most of the shock in straightening itself out. This is particularly true of a properly long tether—one wave length. From that point on, the boat's speed will slow down as the drogue does its job.

A drogue should ride below the surface at such a depth that it is essentially out of the path of the orbiting water particles that make up the wave. To be on the safe side, it would be wise to weight the drogue (or otherwise control the drogue's depth) to about 30 or 40 feet. There is no simple way to measure the depth at which a drogue is operating; however, knowing how long a tether is deployed, and looking at the angle at which it enters the water, an estimate can be made of the drogue's operating depth.

One problem with the cone-type drogue is its tendency to tumble or turn inside out when the pull on the tether is relaxed. This happens because there is significant backwash of water over it when it slows down, which occurs if the tethered boat goes into a wave trough and the drogue does not. When the tether again becomes taut, the cone is snapped back into its streaming position. This continuous snapping of the drogue can place severe strain on the fabric, its stitching, and the tether attachments, leading to early failure unless it was durably built. An open-construction drogue such as the Galerider may not react in this manner.

To alleviate the problem, one has only to load the tether at several places so that there is always some drag along it to keep it stretched out. Donald Jordan suggested the use of a series arrangement of 100 cones, 8 inches in diameter, spliced into 200 feet of line with a 25-pound anchor at the free end (Figure 28). His developmental work on this concept should be closely watched, as it offers advantages in launching, retrieval, and on-board stowage.

Retrieval on a sailboat is simple if you use one of the sheet winches. Simply haul in on the tether until the drogue is at the transom and then manually lift it aboard.

Retrieval for a powerboat without winches is more complicated. If you are in calm waters, you can back the boat up to the drogue, being careful not to get the tether in the propeller or rudder. But the best way is through the use of a trip line as was described for sea anchors. This, of course, has to be rigged before deployment. The method to be used depends on the boat's equipment, so a little forethought will simplify the operation.

OTHER DROGUE USES

Another use for a drogue is to steady the boat when you are forced to run a breaking inlet. Voss demonstrated this several times, using what he called his sea anchor as a drogue. The action is the same as running before a gale. The drogue slows the boat, keeping the stern into the waves—with help from the helmsman. Slowing the boat not only reduces rudder forces but gives the helmsman more time to think and react—things just don't happen so fast when you are going slower.

The only danger with using a drogue for running an inlet is that it can become entangled in rocks, marker buoys, or other hazards surrounding the inlet, especially if the inlet is narrow or has a bend in it. A sharp serrated-blade knife should be kept handy near the transom to sever the tether should a hang-up occur. By that time, however, the boat may well be far enough inside the inlet that it is no longer menaced by the breakers.

Unlike the sea anchor, which takes full control of the boat when deployed, a drogue takes only partial control. The idea of the drogue is to slow the forward progress of the boat (but not to stop it) and to allow the helmsman adequate steering to hold the stern of the boat in a desired position relative to the waves.

Should your boat be anchored in a strong current with an opposing wind coming from the stern, it will have a tendency to stand forward on the anchor rode and do more than a little horsing in the stream. One way to eliminate this problem is to deploy a drogue over the stern, getting added drag to hold the boat firmly against the anchor rode as shown in Figure 33.

A boat being towed will oftentimes tend to run up on the stern of the boat doing the towing, due to either following wind and seas or an inadvertent slowdown of the towing boat. To avoid this, the towed boat can trail a drogue, which will add sufficient drag to prevent its momentum from carrying it forward when the towline slackens. If the disabled boat has lost its rudder, the drogue will also stabilize the towed boat directionally, making it easier to be towed. Generally, the towline in a situation like this is quite short. No attempt is made to have one wave length between tower and towee since such tows would be expected to take place in less than breaking seas.

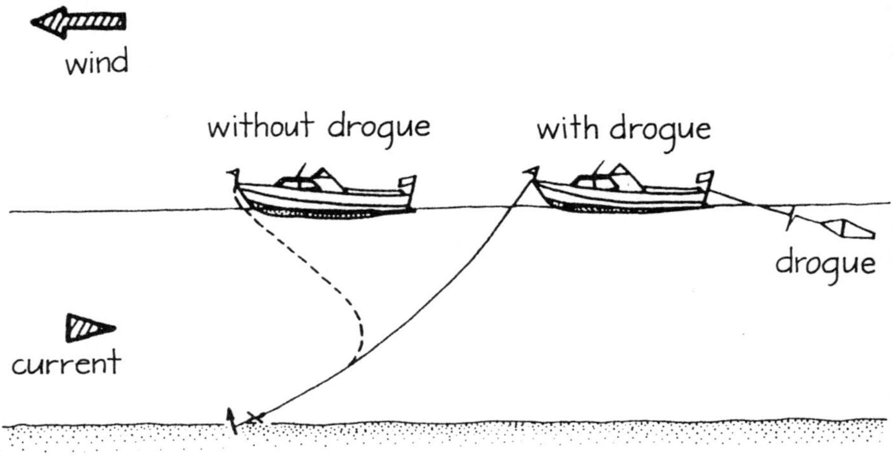

Figure 33. Use of drogue in opposing wind-current situation.

EIGHT

Oil on Troubled Waters

The ability of oil to calm troubled waters is the least studied of all aspects of sea anchoring. Not many people have tried it and there is little unanimity of opinion on how well it works. William Albert Robinson described his experiences (1966) on the 70-foot, 50-ton-displacement brigantine *Varua* when she was caught in the "ultimate storm" in the Great Southern Sea of the Pacific Ocean in 1952.

Varua first hove-to under fore staysail and lower main staysail. Fish oil was pumped through the forward toilet and dispersed from bags hanging to windward to form an oil slick, but the boat kept forereaching "five or six points" off the wind so the oil disappeared to leeward and was of no help. The sails were subsequently lowered and the wheel fixed amidships, causing *Varua* to lie ahull with the seas on her quarter. Now the oil became effective because it formed an upwind slick as *Varua* drifted slowly downwind.

As the storm further developed, however, things got worse. To quote Robinson, "The seas were so huge and concave at this point that the whole upper third seemed to collapse and roar vertically down on us. Our oil had little or no effect now, as the surface water was all blown to leeward." At this point Robinson had to choose another course of action, and this was his thinking:

> It was against prevailing opinion to choose as a last resort to run with the great breaking seas, but I knew instinctively that there would come a point where you could no longer hold her into it, either by drags, sea anchor, riding sails, or any other means. When that point was reached, it would probably be too late to turn and run.

Having tried heaving-to and lying ahull, Robinson felt that the increasing intensity of the storm now demanded that *Varua* run with it. He continues:

> Although under bare poles, Varua picked up speed and began running six or seven knots, dangerously, but steering beautifully. We at once put out drags and slowed her down to about three knots, which still left her with good rudder control. It took a seventy-five fathom, two inch diameter manila line which we dragged in a big bight, plus four seventy-five foot mooring lines of the same size, each dragging its big eye splice, plus about a hundred fathoms of associated lines of smaller size . . . Moving slowly ahead as we now were, we could lay an oil slick right along our path and stern.

The oil slick left behind on the waters with this technique proved quite effective in Robinson's mind.

The quantity of oil dispersed from *Varua* during this storm was quite large and only a vessel the size of *Varua* would have the capacity to carry the required amount. Whereas a small boat may be able to carry sufficient oil for a storm of short duration, it is unlikely that it could carry enough to last several days, which was the case with *Varua.* There is no evidence that a small boat would require any less oil than a large vessel.

The venturesome Capt. Voss used oil when caught in the outskirts of a typhoon in his small boat *Sea Queen* (19 feet LWL). He reported that the oil bags did quite a lot of good in preventing the tops of the seas from breaking off, but as the storm developed further, the oil ceased to help, apparently being blown away or dispersed so much by the turbulent sea that he could see no trace of it on the water.

Another old hand with experience using storm oil was Warwick Tompkins, who brought his schooner *Wanderbird* around the Horn from Gloucester to San Francisco in the early 1930s. Ross Norgrove had met Tompkins and described Tompkins's procedures as follows (Norgrove 1980):

> His method of using oil was simple. He had several dozen bottles, each holding a "fifth" of fish oil, stowed in a locker below decks along with some canvas oil bags. These bags each had a handful of oakum inside (cotton waste will do the same job), and each was capable of holding a bottle of oil. A light line was stitched across the bottom of the bag and up each side, forming a loop at the top; a drawstring closed the mouth.
>
> To use oil in a gale, all he needed to do was open a bottle, push its neck into the mouth of a bag, and let it glug itself empty; no muss, no skidding around in a pool of oil. Then he pulled the drawstring tight, bent a line onto the loop and, with the *Wanderbird* lying ahull, pricked

> the bag with a small spike or icepick and put it over the windward side on 15 or 20 fathoms of the line. For her 80-foot length, this big, engineless schooner used two oil bags (one forward and one aft), and Tompkins stated that they lasted about 17 hours each. He also told me that lying ahull with only these two bags out, he seldom even got his decks wet.

Others have carried storm oil aboard their vessels but never used it. Chichester, Smeeton, and Coles are among those sailors. The logistics of carrying storm oil and having it ready for convenient use are a deterrent to having it on board at all.

The objective of using oil is to increase the surface tension of the sea, thereby reducing the ability of the wind to ruffle the surface. The effect of oil will be greatest on waves in deep water where there is little exchange of water molecules from the surface on down. In shallow waters there is a distinct mixing of surface and subsurface waters, which breaks up the oil film and mixes it with the water.

If waves are not of the mixing variety, only a thin film of oil is required to smooth the surface. In practice, though, there are always some breaking crests that will disperse the oil, requiring that it be replaced in considerable quantity to remain effective.

Sea temperature will also have an effect on the oil dispersal since cold waters will tend to thicken the oil, preventing it from spreading into a film over a large area. Fish and vegetable oils will be affected the most and petroleum oils the least.

The best oil for soothing troubled waters, nevertheless, is fish oil, with heavy vegetable oils (raw linseed oil is very good) and other animal oils next on the list. Heavy petroleum oils such as bunker oil and lubricating oil are of average value. Diesel fuel, kerosene, and gasoline are of no value because they spread too thinly, as can be seen in the harbors of the world.

This need for a special oil to be carried in quantities of tens of gallons makes it unattractive to keep in a small boat's inventory of supplies. In dire circumstances a boat could consider using its spare engine lubricating oils but even these are normally not carried in large enough quantities.

The means for dispersing storm oil is another detractor from the use of oil to calm storm-wracked seas. Voss and others have described the traditional storm oil bag which can be kept empty until ready for use. The bag has a capacity of 3 or 4 pints and can be made of No. 2 canvas or 3-ounce Dacron with securely stitched seams. When it is to be used, the bag is stuffed loosely with oakum (what boat carries oakum on board nowadays?), cotton, kapok, or cotton waste and then filled with the available oil. A number of holes should be punched in the bag with a coarse sail needle or marlinspike to assure adequate seepage but not a heavy flow. The bag can be triced over the windward rail or attached to the sea anchor if one is being deployed.

The dispersal technique used by *Varua* (and a much more practical one) is to pump small quantities of storm oil through a water closet. The rate of flow will have to be arrived at by trial and error and one source recommends a tablespoon every few minutes. Not everyone, though, is going to appreciate the smell of oil belowdecks in a stormy situation when the hatches are closed.

If you have lubricating oil in cans on board, you can punch a couple of small holes in them and hang them one at a time in a mesh bag over the side so each can drip its water-calming contents into the slick created by the hull.

A Coast Guard design for a sea anchor with a built-in dispersal can is shown in Figure 34. The diameter at the mouth is 36 inches and it is approximately 54 inches long with a 1-gallon storm oil container. This would be a very awkward item to stow on board, to say nothing about storing additional storm oil. Incidentally, to replenish the storm oil con-

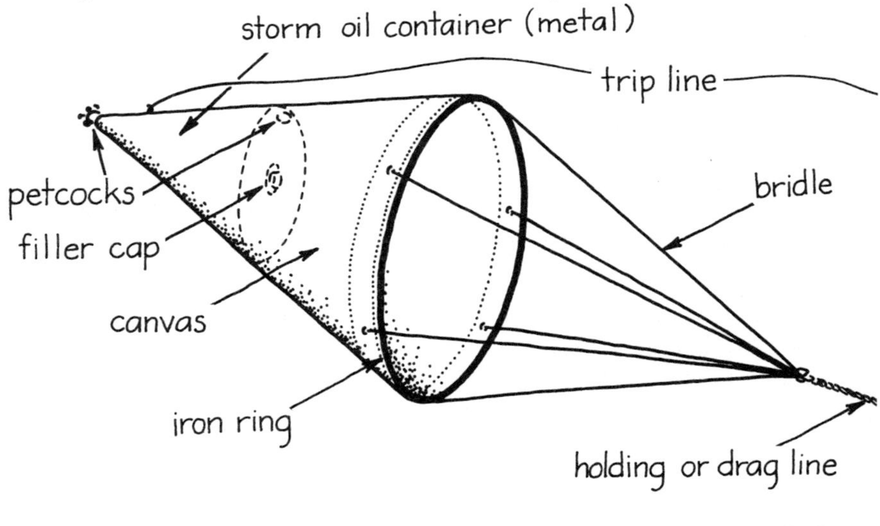

Figure 34. A commercial ship's lifeboat sea anchor. From *U. S. Coast Guard Manual for Lifeboatmen, Able Seamen, and Qualified Members of the Engine Department.*

tainer would require tripping the sea anchor and bringing it back aboard the vessel for refilling and then deploying it once again. Under storm conditions this would seem to be an impractical matter.

Voss noted in his book that there can be other problems with the use of storm oil: Fish oil has a repulsive odor and, if the oil is blown back on the deck or into the cockpit, it could get things dangerously slippery. One should also note that there may be a mess to clean up after the blow; but if you are able to concern yourself with the mess, then you have survived and all is well.

The basic objective in dispersing storm oil is to get it to windward of the struggling vessel and then keep the vessel in the oil slick as much as possible. A rescue vessel to windward of a stricken vessel could dispense oil downwind, hopefully easing the impact of the waves on the stricken vessel, as well as making rescue efforts easier. Onshore waves with an offshore wind could defeat the use of oil in a stranding rescue even though the waves are not breaking heavily. Storm oil dispersal techniques must, obviously, be tailored to the situation and no hard and fast rules can be made.

Finally, one should remember that the discharge of oil at sea from a vessel is in violation of U. S. Coast Guard regulations. However, in an emergency with life and property at risk, it is doubtful whether you should let this technicality stand in your way. The loss of the vessel would free all of the oil in the vessel, causing a somewhat greater ecological and financial calamity.

Epilogue

Can anything with such a history of controversy as drag devices have a future on modern power- and sailboats? Of course they can—provided you are aware of the differences between them. The sea anchor goes over the bow and the drogue goes over the stern. The former stops the boat and the latter slows it down. In the cruel world of storms at sea, misunderstandings of how your equipment functions can be catastrophic.

You should not think of the sea anchor and the drogue as "either/or" devices. Each has its place depending on the sea and the condition of your boat and crew. Even in moderate weather the skipper may simply want to park his boat for awhile and rest, make repairs, or go fishing. To take way off the boat renders it uncontrollable and the boat, regardless of kind, will seek a position parallel to the waves, creating a very uncomfortable ride. A sea anchor will tend to pull the bow of the boat into the wind and make life aboard more pleasant.

Occasionally a boat may encounter a small gale (far from being a survival condition) and, rather than fight it for hours or days on end, the skipper can ride to a sea anchor in relative comfort.

Drogues, however, are rarely used until the storm reaches a near-unsurvivable intensity. Then if the boat is still controllable and the crew capable, it can continue to sail downwind with a drogue preventing it from surfing down the face of a wave and into the bottom of a trough. In the ultimate storm you may elect to turn the boat over to the Jordan series drogue for survival. The options available to the skipper depend on how well he has equipped his boat.

There are a number of drag devices on the market today that, when used with proper seamanship, could give your boat the edge in a survival challenge and their prices are modest by any measure.

Consider the hull insurance on a $100,000 boat (the approximate cost of a 35-foot cruising boat), which would be roughly $3,000 per year.

For about 10 percent of one year's premium you can add either a drogue or a sea anchor (maybe you should carry both) that will give your boat a gale survival capability. Which would you rather have—spare warps, hull insurance, or a dedicated drag device? The answer has to be obvious.

With all the boats sailing the oceans and seas today, we are finding that there is a lot more wild weather than we had expected. A well-prepared blue water boat should be ready to handle all weather and not have to jury-rig some contraption of doubtful performance at the critical moment.

It is a good insurance policy to carry a proven drag device on board your blue water boat. Then, when all else fails, you can at least try it out and form your own conclusion. You have lost nothing and in all probability you have saved the day.

John F. Wilson, master of the SS *Pioneer* observed (Coles 1981): "Whatever decision you make, if you get in trouble, you may be sure that someone who was not there will come up with something you should have done." That something could well have been the timely deployment of a proven drag device.

APPENDIX

The Bottom Line

SEA ANCHOR MANUFACTURERS

1. Para-Anchors International
   P. O. Box 19
   Summerland, CA 93067
   (805) 966-0782

Mouth diameter	*List price**	*Name*
24 feet	$155.00	Para-24
28 feet	$185.00	Para-28
35 feet	$195.00	Para-35
6 feet	$75.00	BuOrd
12 feet	N/A	Force 10

2. Para-Tech Engineering Co.
   10770 Rockville "B"
   Santee, CA 92071
   (619) 440-1664

Mouth diameter	*List price†*
12 feet	$295.00
18 feet	$495.00
24 feet	$795.00
32 feet	$1,295.00

*Without tether or fittings.
†Para-Tech's prices include carrying bag, shackle, 50 feet of float line, and an instruction manual.

3. Shewmon, Inc.
   1000 Harbor Lake Drive
   Safety Harbor, FL 33572
   (813) 447-0091

Mouth diameter	*List price*
4 feet	$149.00
5 feet	$186.00
6 feet	$226.00
8 feet	$322.00
10 feet	$440.00

SPECIALTY SEA ANCHOR MANUFACTURERS

1. Baby sea anchors
   (various manufacturers)

Mouth diameter	*List price*
10 inches	$13.95

2. Blue Harbor, Inc. (Boat Brakes)
   1142 Trailwood North
   Hopkins, MN 55343
   (612) 935-1651

Mouth diameter	*List price*
2 feet	$54.95
2½ feet	$64.95
3 feet	$79.95
3½ feet	$99.95
4 feet	$119.95

3. R. D. Mengler
   10 Hesperia Avenue
   City Beach 6015
   Western Australia

Boat size	*List price*
Up to 20 feet	$119.95
20 to 35 feet	$134.95
35 to 45 feet	$198.95

DROGUE MANUFACTURERS

1. Hathaway, Reiser & Raymond (Galerider)
   184 Selleck Street
   Stamford, CT 06902
   (203) 324-9581

Mouth diameter	*List price*
36 inches	$295.00
41 inches	$349.00

2. Shewmon, Inc. (Variable-Pull Drogue)
   1000 Harbor Lake Drive
   Safety Harbor, FL 33572
   (813) 447-0091

Mouth diameter	*List price*
32 inches	$44.00
54 inches	$60.00
106 inches	$111.00

Glossary

Barber haul: A pendant used to haul a line to one side of its normal direction of pull. May or may not be fitted with a pulley block.

Bridle: A rigging of two or more lines attaching a single object to another line; a means of distributing a load by multiple lines.

Broach: To veer or yaw dangerously, especially in a following sea so as to lie beam-on to the waves.

Canopy: The conical or spherical cloth element of a parachute or a sea anchor.

Capsize: To roll more than 90° in the water. Less than 90° is usually referred to as a knockdown.

Chafe: Wear on a line or piece of fabric caused by rubbing contact with another object.

Chafing gear: Any material used to prevent rope or fabric wearing away when in contact with a solid object.

Crown: The apex of a conical or spherical parachute canopy.

Drogue: A drag device attached to the stern of a boat to slow it down.

Gore: A tapered or triangular segment of a rounded surface.

Lie-to: To keep the bow of a boat into the wind with a sea anchor.

Oil bag: A fabric container for carrying and dispersing storm oil during storm conditions.

Oil slick: An area of water surface covered by a thin film of oil.

Para-anchor: A sea anchor made out of, or like, an airman's parachute.

Pendant: A short or moderate length of line connecting an auxiliary object to a primary piece of gear.

Pitchpole: The most disastrous of boat movements—when a boat digs its bow into the water and goes end over end.

Sea anchor: A drag device attached to the bow of the boat to essentially "anchor" the boat to the surface of the water with the bow held to windward.

Shrouds: The multiplicity of small lines connecting the canopy of a sea anchor or drogue to the tether.

Slick: The smooth water left to windward as a boat moves to leeward.

Spanker: A fore and aft sail set on the sternmost mast of a sailboat (or backstay of a sloop/cutter).

Storm oil: A special oil to increase the surface tension of the water and reduce the ability of the wind to ruffle it.

Tether: The line that connects a sea anchor or drogue to the vessel.

Trip line: A smaller line that connects a sea anchor or drogue apex to the vessel and is used to "trip" or unload the drag device.

Vent: The opening at the apex of a conical or spherical drag device that permits the escape of water and, thereby, stabilizes the device as it is towed through the water.

Bibliography

Bitting, Kenneth R. *The Dynamic Behavior of Nylon and Polyester Line.* Report No. CG-D-39-80. Washington: U. S. Coast Guard, 1980.

Blandford, Percy W. *Modern Sailmaking* (TAB book number 397). Blue Ridge Summit, PA: TAB Books, Inc., 1978.

Bowditch, Nathaniel. *American Practical Navigator.* Pub. No. 9. Washington: Government Printing Office, 1958.

Casanova, John, Joan Casanova, Victor Shane, Daniel C. Shewmon, and Gregory MacMillan. *The Parachute Anchoring System and Other Tactics.* Boston: Chiodi Advertising and Publishing Co., 1982.

Coles, K. Adlard. *Heavy Weather Sailing.* Clinton Corners, NY: John de Graff, Inc., 1981.

Force 10 Imminent. Edition One of Newsletter. Summerland, CA: Para-Anchors International, August 1985.

Gershon, Bradford. *The Mariner's Dictionary.* New York: Weathervane Books, 1972.

Griffith, Bob. *Blue Water—A Guide to Self-Reliant Sailboat Cruising.* Boston: Sail Books, 1979.

Henderson, Richard. *Sea Sense.* Camden, ME: International Marine Publishing Co., 1979.

Hinz, Earl R. *The Complete Book of Anchoring and Mooring.* Centreville, MD: Cornell Maritime Press, Inc., 1986.

Hiscock, Eric C. *Around the World in Wanderer III.* New York: Oxford University Press, 1956.

———. *Cruising under Sail.* New York: Oxford University Press, 1981.

Jordan, Donald J. *Investigation of the Use of Drogues to Improve the Safety of Sailing Yachts and Life Rafts.* Report No. CG-D-37-84. Washington: U. S. Coast Guard, 1984.

Lewis, David. *The Ship Would Not Travel Due West.* N.p.: Temple Press Books, 1961.

Manry, Robert. *Tinkerbelle.* New York: Harper & Row, 1966.

Mate, Ferenc. *The Finely Fitted Yacht* (2 vols.). New York: W. W. Norton, 1979; and Vancouver, B. C.: Albatross Publishing House, 1979.

Moitessier, Bernard. *Cape Horn: The Logical Route.* Bungay, Suffolk: Chaucer Press, 1969.

Norgrove, Ross. *The Cruising Life.* Camden, ME: International Marine Publishing Co., 1980.

Pidgeon, Harry. *Around the World Singlehanded.* London: R. Hart-Davis, 1954.

Poynter, Daniel F. *The Parachute Manual.* North Quincy, MA: Self-published, 1972.

Robinson, William Albert. *To the Great Southern Sea.* London: Peter Davies Ltd., 1966.

Rousmaniere, John. *Fastnet, Force 10.* New York: W. W. Norton, 1980.

Shane, Victor. *Standard Procedures of Parachute Sea-Anchoring.* Summerland, CA: Para-Anchors International, 1983.

Shewmon, Daniel. *Sea Anchors, Drogues and Salvage Bells.* Technical Report. Safety Harbor, FL: Shewmon, Inc., 1981.

United States Yacht Racing Union. *Safety from Capsizing.* Final report in cooperation with the Society of Naval Architects and Marine Engineers. Newport, RI: USYRU, 1985.

Van Dorn, William G. *Oceanography and Seamanship.* New York: Dodd, Mead & Co., 1974.

Voss, John Claus. *The Venturesome Voyages of Captain Voss.* Sydney, B. C.: Gray's Publishing Ltd., 1949.

Walton-Smith, F. G. *The Seas in Motion.* New York: Thomas Y. Crowell Co., 1973

Index

About the Author

Earl R. Hinz, an inveterate sailor (over 40,000 miles logged) and former aeronautical engineer, retired in 1975 to devote his time to sailing and to writing about boats and boating. From 1976 to 1981 he was on the staff of *Sea* as technical editor, writing pioneering articles on instrumentation and techniques for evaluating the performance of power- and sailboats. He personally conducted sea trials on over 100 recreational vessels. In 1979 he won the prestigious Boating Writer's International Award for his magazine report on a test program he conducted on boating equipment. In 1980 he started a monthly column on Pacific cruising for *Sea;* this column later became a standard bimonthly feature of *Cruising World.*

Now a resident of Honolulu, Mr. Hinz is also the author of *The Offshore Log, Sail Before Sunset, Landfalls of Paradise,* and *The Complete Book of Anchoring and Mooring.*

Richard R. Rhodes is a free-lance graphic designer and technical artist specializing in geophysical and marine illustrations. An avid sailor himself, he was responsible for the design and specifications of the Hawaiian Koa Racing Canoe. Most recently he was part of the crew of *Hokule'a,* the Polynesian voyaging canoe, on its return from New Zealand.